Pearson Australia
(a division of Pearson Australia Group Pty Ltd)
459–471 Church Street, Level 1, Building B,
Richmond, Victoria 3121
PO Box 23360, Melbourne, Victoria 8012
www.pearson.com.au

First published 2017 by Pearson Australia
2026 2025 2024 2023
10 9 8 7 6 5 4 3 2 1

Publisher: Alicia Brown
Development Editor: Shirley Melissas
Project Manager: Shelly Wang
Production Manager: Elizabeth Gosman
Editors: Deborah Lombardo, Aptara
Proof Reader: Vicky Chadfield
Designer: Anne Donald
Copyright & Pictures Editor: Samantha Russell-Tulip
Desktop Operator: Anita Adams and Aptara
Illustrator: DiacriTech
Printed in Malaysia (CTP-VVP)

ISBN 9781488615054

Pearson Australia Group Pty Ltd ABN 40 004 245 943

Disclaimer/s
The selection of internet addresses (URLs) provided for this book/resource was valid at the time of publication and was chosen as being appropriate for use as a secondary education research tool. However, due to the dynamic nature of the internet, some addresses may have changed, may have ceased to exist since publication, or may inadvertently link to sites with content that could be considered offensive or inappropriate. While the authors and publisher regret any inconvenience this may cause readers, no responsibility for any such changes or unforeseeable errors can be accepted by either the authors or the publisher.

Some of the images used in *Pearson Science* might have associations with deceased Indigenous Australians. Please be aware that these images might cause sadness or distress in Aboriginal or Torres Strait Islander communities.

Practical activities:
All practical activities, including the illustrations, are provided as a guide only and the accuracy of such information cannot be guaranteed. Teachers must assess the appropriateness of an activity and take into account the experience of their students and facilities available. Additionally, all practical activities should be trialled before they are attempted with students and a risk assessment must be completed. All care should be taken and appropriate protective clothing and equipment should be worn when carrying out any practical activity. Although all practical activities have been written with safety in mind, Pearson Australia and the authors do not accept any responsibility for the information contained in or relating to the practical activities, and are not liable for any loss and/or injury arising from or sustained as a result of conducting any of the practical activities described in this book.

Acknowledgements
We would like to thank the following for permission to reproduce copyright material. The following abbreviations are used in this list: t = top, b = bottom, l= left, r = right, c = centre.

Cover image: Alamy Stock Photo/WaterFrame

123RF: pp.16 (ice), 38 (pump nozzle); Elena Duvernay, p. 16 (rain); Don Fink, p. 70; Milosh Kojadinovich, p. 16 (jelly); Tamara Kulikova, p. 16 (smoke); lightwise, p. 96; Serdar çorbacı, p. 37; server, p. 16 (timber); Leah-Anne Thompson, p. 16 (toothpaste)

ABC: 'Sand playground makes for safer landings' by Annabel McGilvray, 15 December 2009. Reproduced with permission of the Australian Broadcasting Corporation—Library Sales © 2009, p. 107

Alamy Stock Photo: NASA Archive, p. 119; Genevieve Vallee, p. 50t

Fairfax Photo Sales: The Sydney Morning Herald, p. 94bl

Getty Images: Time Life Pictures, p. 26

Daniel Solander Library, Royal Botanic Gardens and Domains Trust/Linnean Society of NSW: p. 94tr

Office of Environment & Heritage, NSW: © State of New South Wales and Office of Environment and Heritage 2009. Licensed under Creative Commons (CC BY 4.0), p. 45

Pearson Education Ltd: Studio 8, p. 16 (slime)

Shutterstock: 673486, p. 38 (algae); Jeff Banke, p. 50c; Jim Barber, p. 38 (soybeans); Ivan Cholakov, p. 38 (microorganism); Le Do, p. 92bl; David Dohnal, p. 91t; David Good, p. 91bl; Free-lance, p. 91cr; foxbat, p. 33cl; irin-k, p. 92t; Klaus Kaulitzki, p. 33r; Jesus Keller, p. 33l; marekuliasz, p. 62; Michelle D. Milliman, p. 16 (bubblegum); Christian Musat, p. 91cl; pzAxe, p. 92br; Zeljko Radojko, p. 38 (corn field); sanneberg, p. 50b; Audrey Snider-Bell, p. 92bc; three one five, p.16 (aerosol); TranceDrumer, p. 33cr; vblinov, p. 92c; Denis Vrublevski, p. 16 (fruit); wavebreakmedia, p. xiii; XAOC, p. 91br

Skymaps.com: Kym Thalassoudis, Skymaps.com, p. 120

Thinkstock: jarnogz, p. 43; Jupiterimages, p. 78

Every effort has been made to trace and acknowledge copyright. However, should any infringement have occurred, the publishers tender their apologies and invite copyright owners to contact them.

Contents

Pearson Science 2nd edition Activity Book

An intuitive, self-paced approach to science education, which ensures every student has opportunities to practise, apply and extend their learning through a range of supportive and challenging activities.

Pearson Science 2nd edition has been updated to fully address all strands of the new Australian Curriculum: Science, which has been adopted throughout the nation. This edition also captures the coverage of Science curricula in states such as Victoria, which have tailored the Australian Curriculum slightly for their students.

The *Pearson Science 2nd edition* features a more explicit coverage of the curriculum. The activities enable flexibility in the approach to teaching and learning. There are opportunities for extension as well as reinforcement of key concepts and knowledge. Students are also guided in self-reflection at the end of each topic.

Explicit scaffolding makes learning objectives clear and includes regular opportunities for reflection and self-evaluation.

In this edition, we provide a structured approach that integrates a seamless, intuitive and research-based learning hence **differentiating** the course for every student.

The Activity Book also provides richer application opportunities to take the Student Book content further with explicit coverage of Inquiry Skills, Science as a Human Endeavour and Science Understanding.

The diverse offering of worksheets allows students to be challenged at their level. Students have the flexibility to be self-paced and this new edition comes with the advantage of each worksheet being self-contained.

Be guided

A new handy **Toolkit** at the beginning of the Activity Book has been created to build skills in the key areas of practical investigations, research, thinking, organising, collecting and presenting. Each skill developed in the toolkit is directly relevant to applications in questions, investigations and research activities throughout the student and activity books. A toolkit spread provides guides and checklists alongside models and exemplars.

Be supported

Vocabulary boxes provide definitions for key terms, within the relevant context of the task. **Hints** help students get started on a worksheet and provide support in overcoming a barrier.

Be reflective

The **Thinking about my learning** feature provides the opportunity for self-reflection and self-assessment. It encourages students to look ahead to how they can continue to improve and assists in highlighting focus areas for skill and knowledge development.

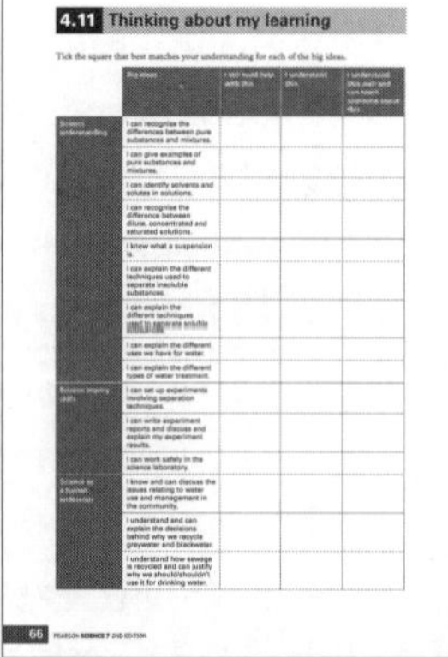

Be ready

A **knowledge preview** at the beginning of every chapter, activates prior knowledge relevant to the topic, providing an opportunity for students to show what they currently know. This handy tool supports teachers in assessing students' prior knowledge.

Be literate

Newly improved **literacy reviews**, in consultation with our Literacy Consultant Dr Trish Weekes, provide a deeper and broader range of language building tasks. Every chapter concludes with a literacy review which focuses on building a deeper understanding of key terms supporting students to correctly apply key terms from the topic.

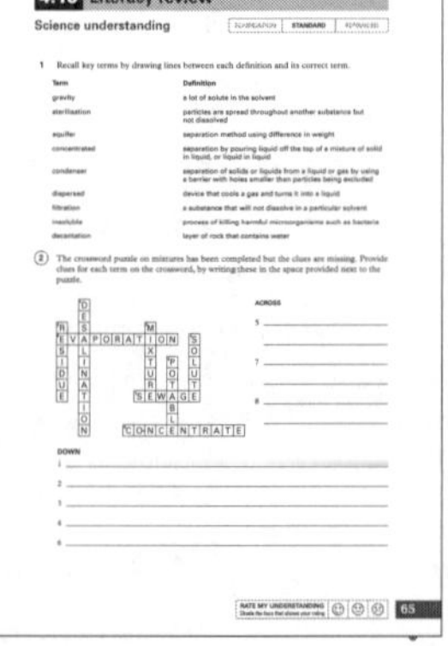

Be set

Visit **www.pearsonplaces.com.au** for digital assets and interactive resources:

- Interactive activities and lessons
- Untamed Science videos
- Weblinks
- Student investigation templates and teacher support
- Answers for Activity Book and Student Book questions and tests
- SPARKlabs
- Risk assessments
- Teaching programs and curriculum mapping audits

Worksheets

Each worksheet is classified according to the degree with which it deals with curriculum understandings.

- **Foundation** indicates the focus is on the basics like terminology.
- **Standard** indicates a focus on the core ideas, understandings and skills.
- **Advanced** indicates transfer and extension of core science understanding and skills to new or more sophisticated situations.

Teachers may use this tool to **differentiate** the worksheets allocated to students. They may select worksheets for students based on whether basic, core or extension exercises are required. The categories do not indicate the degree of difficulty of tasks on the worksheet. A worksheet labelled advanced may have tasks ranging from lower level through to higher-level thinking.

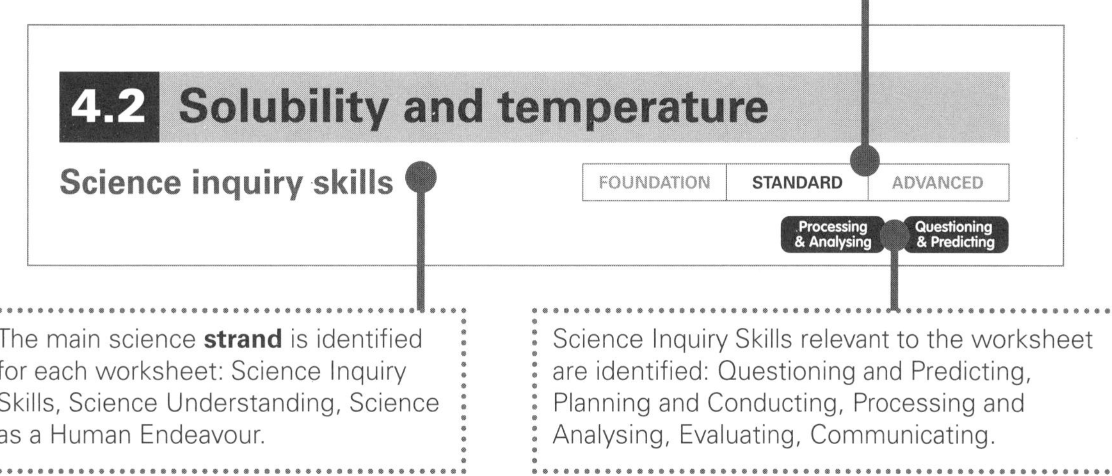

The main science **strand** is identified for each worksheet: Science Inquiry Skills, Science Understanding, Science as a Human Endeavour.

Science Inquiry Skills relevant to the worksheet are identified: Questioning and Predicting, Planning and Conducting, Processing and Analysing, Evaluating, Communicating.

Each question in a worksheet is identified according to the degree of difficulty. Bloom's taxonomy is used as the basis for question classification. The classification is intentionally subtle and unobtrusive. This tool may be used to differentiate between students, matching questions to their levels of ability.

Questions with a straight number indicate a remembering or understanding lower-order question.

Questions with a circle around the question number indicate an analysing or applying middle-order question.

Questions with a square around the question number indicate a creating or evaluating higher-order question.

1 Read through the words in the vocabulary

(a) List the words that you know and wr

(5) Identify the solutes that did not dissolve in

[7] Discuss how this story illustrates that scien

An innovative tool for students to quickly and easily reflect on their understanding of each worksheet. The teacher may use the student responses as a formative assessment tool. At a glance, teachers may assess which topics and which students need intervention for improvement.

RATE MY UNDERSTANDING
Shade the face that shows your rating   

PEARSON Australian Curriculum Writing and Development Team

Anna Bennett
Differentiation consultant, Victoria

Ian Bentley
Former Head of Science, VCE exam and trial exam writer, STEM investigation developer, Victoria

Christina Bliss
Science and senior Biology teacher, VCE assessor, Biology author and question writer, Victoria

Donna Chapman
Science laboratory technician, Safety consultant, Victoria

Dr Warrick Clarke
Curriculum writer, Science communicator and Australian Post-Doctoral Research Fellow at UNSW, author, New South Wales

Jacinta Devlin
Science and senior Physics teacher, author, Victoria

Julia Ferguson
Education Officer Earth Science Western Australia, author and reviewer, Western Australia

Bob Hoogendoorn
VCAA exam assessor in Chemistry, former senior Chemistry teacher exam-style question author, Victoria

Penny Lee
Science laboratory technician, safety consultant, Victoria

Louise Lennard
Head of Science, former industrial scientist, author and STEM investigation developer, Victoria

Greg Linstead
Former Head of Science, Education Department of WA curriculum writer, Deputy Chief Examiner, STAWA President, author, Western Australia

Bryony Lowe
Director of Numeracy Improvement. Former Head of Science, Region Teaching & Learning Coach, author and reviewer, Victoria

David Madden
Science Learning Area Manager at QCAA, former Head of Science, author and reviewer, Queensland

Fran Maher
Bioscience educator, Science teacher, former Bioscience researcher author, Victoria

Rochelle Manners
Science and Mathematics teacher, Teacher Companion coordinating author, Queensland

Shirley Melissas
Teacher librarian and author, Development editor, Victoria

Tamsin Moore
Science and senior Psychology teacher, author, Western Australia

Natalie Nejad
Head of Science, Science and Mathematics teacher, STEM investigation developer, Victoria

Malcolm Parsons
Education consultant, former teacher, author, Victoria

Greg Rickard
Teacher, former Head of Science, author, Victoria

Lana Salfinger
Teacher, Head of Science, IB Workshop leader in MYP Sciences author, Western Australia

Maggie Spenceley
Former teacher, curriculum writer Queensland Studies Authority, author, Queensland

Jim Sturgiss
Teacher, former coordinating analyst (NSW Department of Education) reporting NAPLAN, senior test designer for Essential Secondary Science Assessment (ESSA), Thinking Scientifically questions author, New South Wales

Craig Tilley
STAV trial exam coordinator, VCAA exam assessor, former senior Physics teacher, author and exam-style question writer, Victoria

Jo Watkins
Chief Executive Officer at Earth Science Western Australia, author and reviewer, Western Australia

Dr Trish Weekes
Science literacy consultant, New South Wales

Rebecca Wood
Science educator and tutor, author, Victoria

Science toolkit

Thinking scientifically

Scientific writing

The 'voice' refers to the style of writing. Science writing has a distinct voice or style.

Factual

Scientific writing is factual. This means that it provides information or facts that are known to be true. Factual writing does not use emotions or feelings when describing or explaining things. This makes the writing more neutral and impersonal, as shown in this comparison.

✔ Sewage, which is waste water from kitchens and bathrooms, can be recycled to make potable water suitable for human consumption.

✘ Sewage which is waste water from kitchens and bathrooms can be recycled but it is disgusting and I would not drink it.

Formal

Scientific writing does not use contractions such as 'don't'. Instead, spell out words fully. Use 'do not' instead of 'don't'. Colloquialisms used in conversation should be replaced by formal language. Notice the formal style of the first sentence in this example.

✔ Children were told the risks of using the Bunsen burner and importance of using exactly 5cm of magnesium ribbon for the experiment.

✘ The kids were told to watch it using the Bunsen burner and to get about 5cm of magnesium ribbon for the experiment.

Passive

The passive voice uses the third person rather than the first person. Things rather than people are the subjects of sentences. This means that the active voice and words 'I', 'me', 'my', 'we' or 'mine' are not used, as shown below.

✔ It was noticed that the solution turned yellow when heated.
✔ A line graph was used to show the results.

✘ I noticed that the solution turned yellow when I heated it.
✘ We used a line graph to show the results.

Scientific visuals

Scientific diagrams are used to present written information in a more simplified, visual way.

Labelled diagrams and drawings

Diagrams and drawings aid understanding of information and ideas, and can be used to represent concepts that cannot be seen or are difficult to understand. Drawings provide a lot of information in a concise way that is quick and easy to understand. Figures 0.1, 0.2 and 0.3 summarise features of scientific diagrams and drawings.

Features of scientific drawings in general

Features of scientific drawings and diagrams		
General drawings and diagrams	• are accurate representations reflecting shape, colour and proportions.	• labels, notes or a key can describe and identify features. • arrows should only be used to show processes, not to label parts and features.
Experiment report drawings	• are two-dimensional (2D) with carefully ruled lines, in correct proportions. • equipment is labelled and connected to the object with a pointer line.	• labels are written in print, not running writing, on handwritten reports.

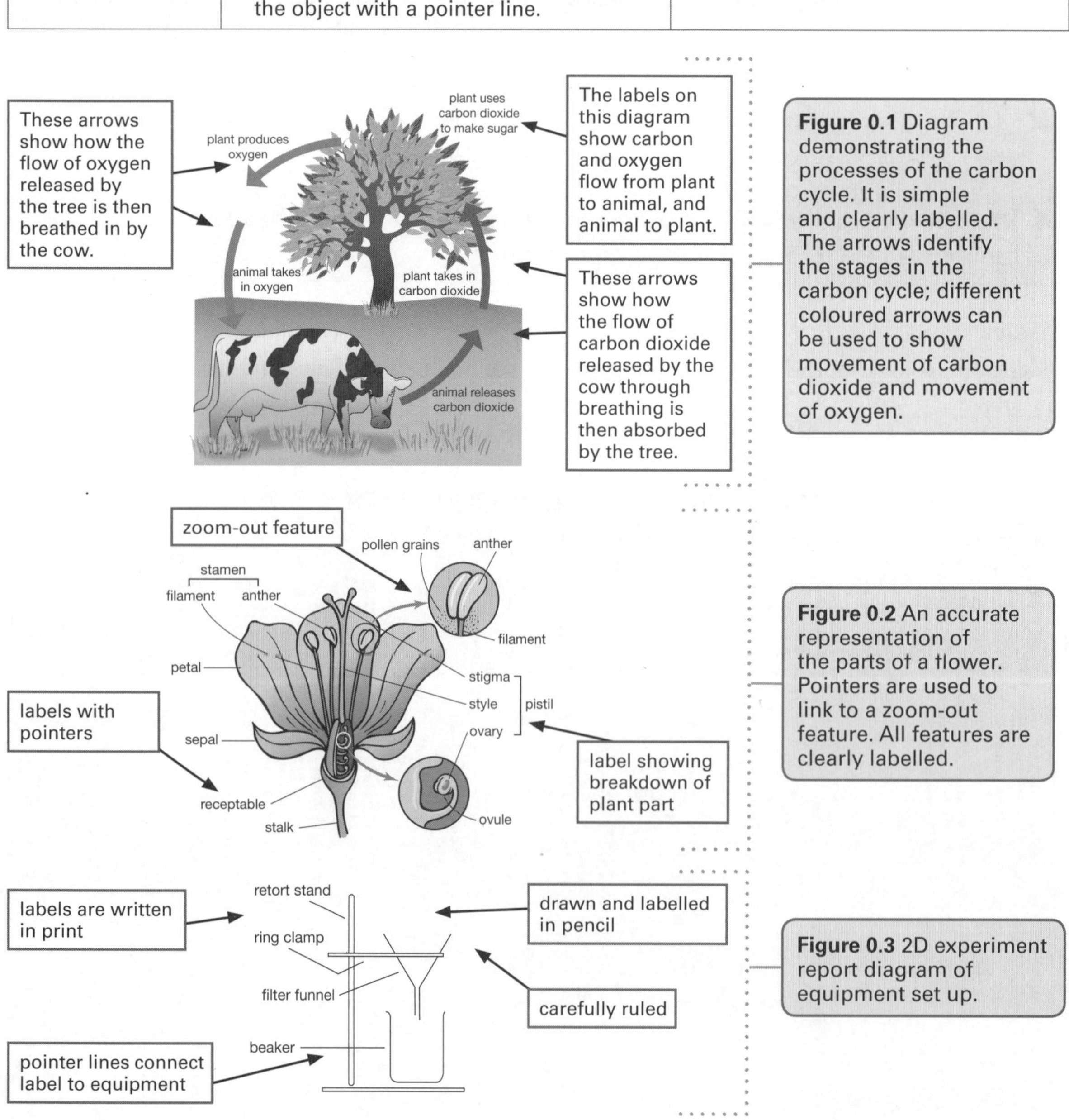

Figure 0.1 Diagram demonstrating the processes of the carbon cycle. It is simple and clearly labelled. The arrows identify the stages in the carbon cycle; different coloured arrows can be used to show movement of carbon dioxide and movement of oxygen.

Figure 0.2 An accurate representation of the parts of a flower. Pointers are used to link to a zoom-out feature. All features are clearly labelled.

Figure 0.3 2D experiment report diagram of equipment set up.

Science toolkit

Graphic organisers

Graphic organisers are charts or diagrams that are used to help organise and present information or ideas. Graphic organisers can be very helpful for visual learners. There are a number of commonly used graphic organisers.

Flow charts

A flow chart uses boxes and words, as well as images, to demonstrate the sequence of events in a process or the way things are connected with each other (Figure 0.4).

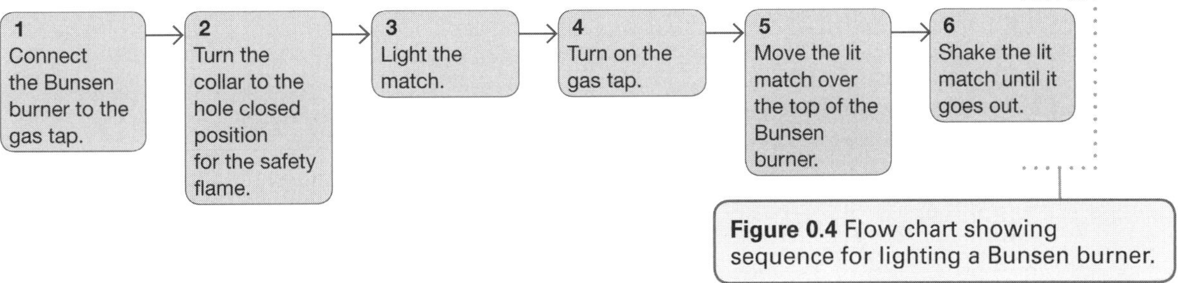

Figure 0.4 Flow chart showing sequence for lighting a Bunsen burner.

Concept maps

A concept map organises ideas in a hierarchical branching structure. It uses words and captions. Ideas are linked by arrows. Ideas may also be linked with phrases like 'leads to', 'results in' and 'impacts on'. The concept map below looks at the issue of water management, water shortage and the causes and relationships between them.

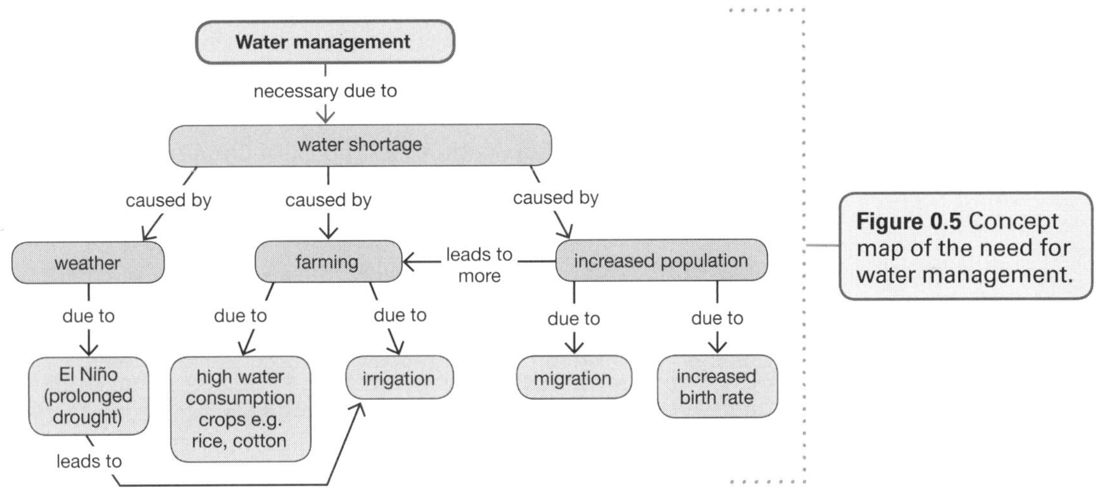

Figure 0.5 Concept map of the need for water management.

Venn diagrams

A Venn diagram can be used to compare and contrast two items. The overlap section lists common features of the two items. The outer circles list the unique features of each item (Figure 0.6).

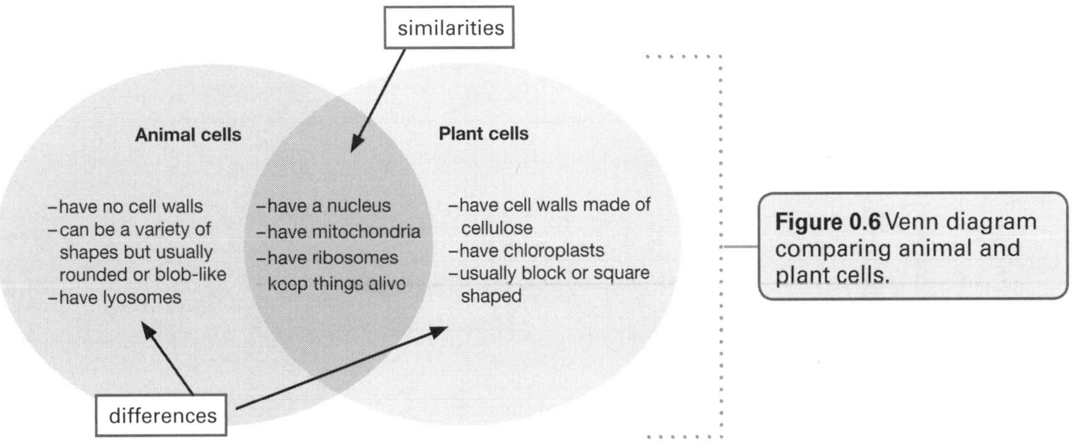

Figure 0.6 Venn diagram comparing animal and plant cells.

Working with scientific data

Science involves the collection and analysis of data. Data often comes in the form of numbers. Number data is quantitative information such as changes in temperature, the height and width of an object or how much liquid an object can hold. Data can be presented in many different ways including in tables and graphs.

Tables

Tables are a clear and simple way to record information. Each table column and row must have a heading that lists the items, characteristics or variables that are being compared. The table must have a title that explains what information is being presented (Figure 0.7).

Substances in sand and their characteristics

Substances in sand	Characteristics of substances		
	Magnetic	Weight	Fragment size
silica	no	not heavy	large
coral and shell fragments	no	not heavy	large
feldspar	no	not heavy	small
magnetite	no	very heavy	small
rutile	no	very heavy	small

Figure 0.7 Table of the characteristics of each substance in sand.

Graphs

A graph is used to show the relationship between two or more variables. For example, a graph might show the average winter temperatures of each Australian state and territory. There are different types of graphs—bar or column, pie and line—but all graphs have certain features in common. A helpful way to remember these features is by using the acronym SALTS—scale, axes, labels, title, source.

SALTS: common features of graphs

Scale	Axes	Labels	Title	Source
numbers (in even increments) and/or words that show the units/items used on a graph	the *x*-axis runs horizontally the *y*-axis runs vertically	axes need to be labelled with the name of the variable they show	explains what the graph is about or what the data displays	some graphs also include the source, identifying where data comes from

Column or bar graphs

Column and bar graphs are similar. The main difference is that column graphs have vertical bars on the *x*-axis and bar graphs have horizontal bars on the *y*-axis as shown in Figure 0.8. The data in these types of graphs are not related so one value does not depend on the next. This data is called discrete data—data that can be sorted into groups and counted.

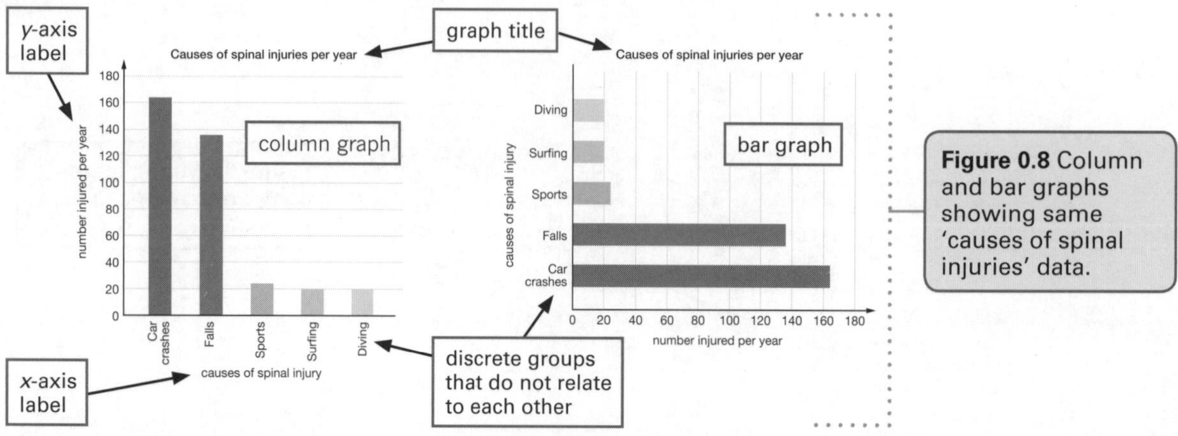

Figure 0.8 Column and bar graphs showing same 'causes of spinal injuries' data.

Line graphs

Line graphs are a special type of scatter plot. This type of graph compares two variables that are related to one another. Line graphs often show changes in data over time intervals. Data is plotted as points, then points are joined to make a line. Continuous data is represented as a line graph. Continuous data means you can choose a point between two plotted graph points and then work out the measurement of the variable at that point. For example, Figure 0.9 shows that the tea's temperature is measured every 1 minute (as shown on the *x*-axis). From the line of plotted points, you can work out that 30 seconds after the tea is made the temperature is 35° Celsius.

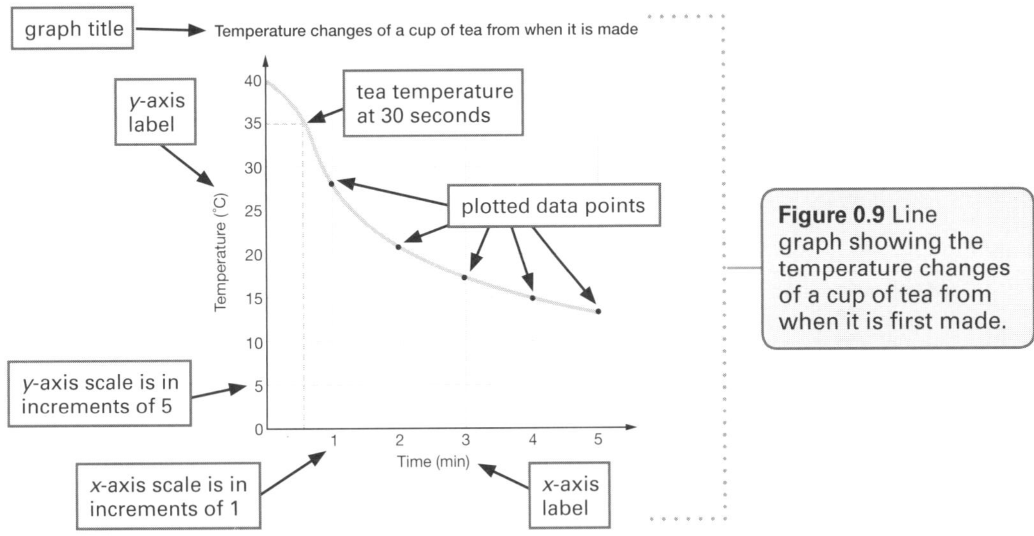

Figure 0.9 Line graph showing the temperature changes of a cup of tea from when it is first made.

Scientific practical reports

Scientific practical reports are records of the all aspects of an experiment. You will be asked to write a practical report after completing an experiment in class. Practical reports have a clear organisation and structure, as shown in Figure 0.10. Figure 0.11 is a model of a well-written practical report, while the model report in Figure 0.12 needs a little improvement.

Practical report structure

Title and date: The title explains what the experiment is about.

Purpose: This explains the aim of your experiment—what you were hoping to achieve or demonstrate. The purpose usually starts with 'To' (for example, 'To demonstrate that...' or 'To test whether...').

Hypothesis: A statement of the expected result of the experiment, expressed as 'If ... then ...'.

Materials: This is a list of all the equipment, chemicals and/or materials (including sizes and amounts) required to complete the experiment.

Procedure/Method: A detailed explanation of what you do in the experiment. Number each step of the process and provide exact quantities (for example, 5 g of copper sulfate) and techniques used so that the reader could repeat the experiment in exactly the same way. A labelled diagram can help explain the procedure.

Results: A summary of what happened in the experiment. The data is often presented as a table or graph.

Discussion: Provide an explanation of what you think the results of the experiment demonstrate. You may also discuss any problems you encountered and/or suggestions for improvement.

Conclusion: A short restatement of the results of the experiment and how they link back to the hypothesis of the experiment.

Figure 0.10

Science toolkit

Model of a well-written practical report for a simple experiment

16th May 2016

COMBINING OIL AND WATER

Includes a clear title that outlines the purpose of the experiment

Purpose: To determine whether oil and water can be mixed together.

Clearly describes what the experiment aims to achieve

Hypothesis: If oil is mixed with water then the oil will remain separated from the water.

A prediction is made about the result of the investigation.

Materials

150 mL of water
150 mL of vegetable oil
3 glass beakers
3 spatulas or stirring rods
2 x 250 mL measuring cylinders
stopwatch or digital recorder

Provides a list of quantities and types of materials required

Procedure

1. Measure out 50 mL water using a measuring cylinder.
2. In a second measuring cylinder, measure out 50 mL of oil.
3. Pour 50 mL water into one beaker.
4. Slowly pour 50 mL of oil into the same beaker, on top of the water.
5. Observe what happens and record.
6. Mix oil and water together using a stirring rod for 30 seconds.
7. Observe what happens and record.
8. After 1 minute, observe the beaker with oil and water and record.
9. Repeat steps 1 to 4 two more times. Make sure that with each trial clean beakers and stirring rods are used as well as new water and oil.

Steps are numbered and written clearly and simply as instructions.

Precise instructions are given, including length of time and manner in which mixture was stirred.

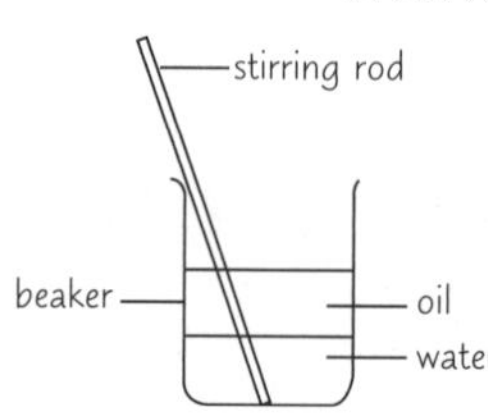

A 2D scientific diagram is correctly drawn and demonstrates equipment and method used.

Results

Observations before, during and after oil was mixed with water

Trial	One minute before mixing	While being mixed	One minute after mixing
1	oil and water separated with oil sitting on top of water	oil and water swirled through each other with oil forming globules in the water	oil and water separated with oil sitting on top of water
2	oil and water separated with oil sitting on top of water	oil and water swirled through each other with oil forming globules in the water	oil and water separated with oil sitting on top of water
3	oil and water separated with oil sitting on top of water	oil and water swirled through each other with oil forming globules in the water	oil and water separated with oil sitting on top of water

When 50 mL of oil was first added to 50 mL of water in the beaker, the two substances separated and the oil sat on top of the water. When vigorously mixed for 30 seconds, the oil broke up into globules that mixed around in the water, but the two substances did not become one solution. One minute after stirring stopped, the oil separated and formed a layer again, on top of the water.

A table of results is provided. As these are qualitative rather than quantitative, a graph was not used to display results.

There is a brief written summary of results.

Discussion

Oil and water do not mix because water molecules are attracted to each other rather than to the oil molecules. The oil molecules remain at the top because they are less dense than water molecules.

An explanation of the results is provided.

Try to link findings to the theory.

Make suggestions for improvement.

Conclusion

Oil and water cannot be successfully mixed together, even when they are very vigorously stirred.

The conclusion is clear and short.

It relates back to the hypothesis.

Figure 0.11

Model of a practical report that needs improvement

Title should be written as a statement, not as a question.

There is no date indicating when the experiment took place.

CAN OIL AND WATER MIX?

Purpose: I wanted to see if I could mix the two substances together.

There is a purpose but no hypothesis for this experiment.

Do not use 'I' in a practical report.

Be specific by naming the substances.

Materials

Some water, some oil, beakers, spoon, rod, syringe or dropper, clock

Quantities of materials and substances must be listed. The specific type of oil should be named. Unnecessary items should not be listed. Use scientific equipment with measurement scale for accuracy.

Procedure

Mix the oil and water together and see what happens.

Steps in procedure need to be numbered and the directions detailed so the procedure can be repeated exactly by someone else.

A labelled diagram would assist in explaining the procedure.

The procedure lacks details such as the stirring of the mixture, the quantities of substances combined, how many trials there were.

Results

When the oil was added into the beaker I did not see anything happen so I mixed it around a bit but still nothing. I wanted it to explode or something!

Keep language formal and make sure the results are clearly explained. Write in a passive voice, not using 'I'.

Avoid emotional language.

Be exact about how long the mixture was stirred.

Discussion

I'm not sure why the two substances didn't mix together.

The discussion must provide an explanation of the experiment's results.

Do not use abbreviations like 'didn't'.

The report does not have a conclusion.

Figure 0.12

Risk assessments

Science investigations should be completed with care to make sure you are safe at all times. There are potential risks with some investigations, so these need to be considered and managed.

Before starting any investigation, make sure you understand the safety rules and possible risks so you can reduce the chance of something going wrong and someone becoming injured.

The risk assessment form (Figure 0.13) is a useful tool to help you identify possible risks and to think about the measures you can take to reduce them.

Science students stay safe by wearing goggles and protective laboratory coats during a chemistry experiment.

Risk assessment form	
Title or description of the investigation	
Names of students in the group	
Class:	Date:
What are the RISKS involved?	**How can you REDUCE these risks?**
List the equipment you will be using. ________ ________ ________ ________	State how you will safely use each piece of equipment. ________ ________ ________ ________
List the chemicals you will be using. ________ ________ ________ ________ ________	State how you will carefully use each chemical. ________ ________ State how you will safely dispose of chemicals. ________ ________
List any other possible risks. ________ ________ ________	State how each risk will be reduced. ________ ________ ________

Figure 0.13

Learning styles

Not all people learn in the same way. Some people like to read information while others learn best when they hear information. The way (or ways) that you learn most effectively is known as your preferred learning style. No one style is better than another.

The three main learning styles are:

Visual
These learners prefer to learn by seeing and observing things. Information presented as images, diagrams, maps, pictures, photos or films suits visual learners.

Auditory
These learners prefer to learn through listening. This could be oral information from the teacher, speeches or group discussions, documentary films.

Kinaesthetic
These learners prefer to learn through 'doing' rather than listening or seeing. Some examples include making models, conducting experiments or role-playing situations.

Some people may have one main preferred learning style whereas others may have a combination of two or even all three. Take the quiz below to discover your preferred learning style or styles. Circle one preferred option (A, B or C) for each sentence stem.

Learning style quiz	
If I am teaching someone something new, I tend to: A write instructions down for them B give them a verbal explanation C demonstrate first and then let them have a go	**I feel especially connected to other people because of:** A how they look B what they say to me C how they make me feel
If I am explaining to someone, I tend to: A show them what I mean B explain to them in different ways until they understand C encourage them to try and talk them through my ideas as they do it	**When I concentrate, I most often:** A focus on the words or pictures in front of me B discuss the problem and the possible solutions in my head C move around a lot, fiddle with pens and pencils and touch things
During my free time, I most enjoy: A going to museums and galleries B listening to music and talking to my friends C playing sport or doing DIY	**My first memory is of:** A looking at something B being spoken to C doing something
When I go shopping for clothes, I tend to: A imagine what they would look like on B discuss them with the shop staff C try them on and test them out	**When I am anxious, I:** A visualise the worst-case scenarios B talk over in my head what worries me most C can't sit still, fiddle and move around constantly
When I am learning a new skill, I am most comfortable: A watching what the teacher is doing B talking through with the teacher exactly what I am supposed to do C giving it a try myself and working it out as I go	**When I listen to a band, I can't help:** A watching the band members and other people in the audience B listening to the lyrics and the beat C moving in time with the music

Science toolkit

Learning style quiz	
If I am choosing food off a menu, I tend to: A imagine what the food will look like B talk through the options in my head or with my family C imagine what the food will taste like	**I tend to say:** A watch how I do it B listen to me explain C you have a go
I first notice how people: A look and dress B sound and speak C stand and move	**I think that you can tell someone is lying if:** A they avoid looking at you B their voice changes C they give me funny vibes
Most of my free time is spent: A watching television B talking to friends C doing physical activities or making things	**When I meet an old friend:** A I say 'it's great to see you!' B I say 'it's great to hear from you!' C I give them a hug or a handshake
I really love: A watching films, photography, looking at art or people watching B listening to music, the radio or talking to friends C taking part in sporting activities, eating yummy foods or dancing	**I remember things best by:** A writing notes or keeping printed details B saying them aloud or repeating words and key points in my head C doing and practising the activity or imagining it being done
If I am angry, I tend to: A keep replaying in my mind what it is that has upset me B raise my voice and tell people how I feel C stamp about, slam doors and physically demonstrate my anger	**When I have to revise for a test, I generally:** A write lots of revision notes and diagrams B talk over my notes alone or with friends C imagine making the movement or creating the formula
I find it easiest to remember: A faces B names C things I have done	**I tend to say:** A I see what you mean B I hear what you are saying C I know how you feel

Now add up how many As, Bs and Cs you chose and write the total of each in the boxes below.

Number of As ___________

Number of Bs ___________

Number of Cs ___________

If you chose mostly As, you prefer a visual learning style. ☐
If you chose mostly Bs, you prefer an auditory learning style. ☐
If you chose mostly Cs, you prefer a kinaesthetic learning style. ☐

Now that you are aware of your preferred learning style or styles, reread the descriptions of those style/s.

CHAPTER 1

Working scientifically

1.1 Knowledge preview

Science understanding

FOUNDATION | STANDARD | ADVANCED

1 Work with a partner or in groups of three. Think about what you know about science and complete the Y-chart below.

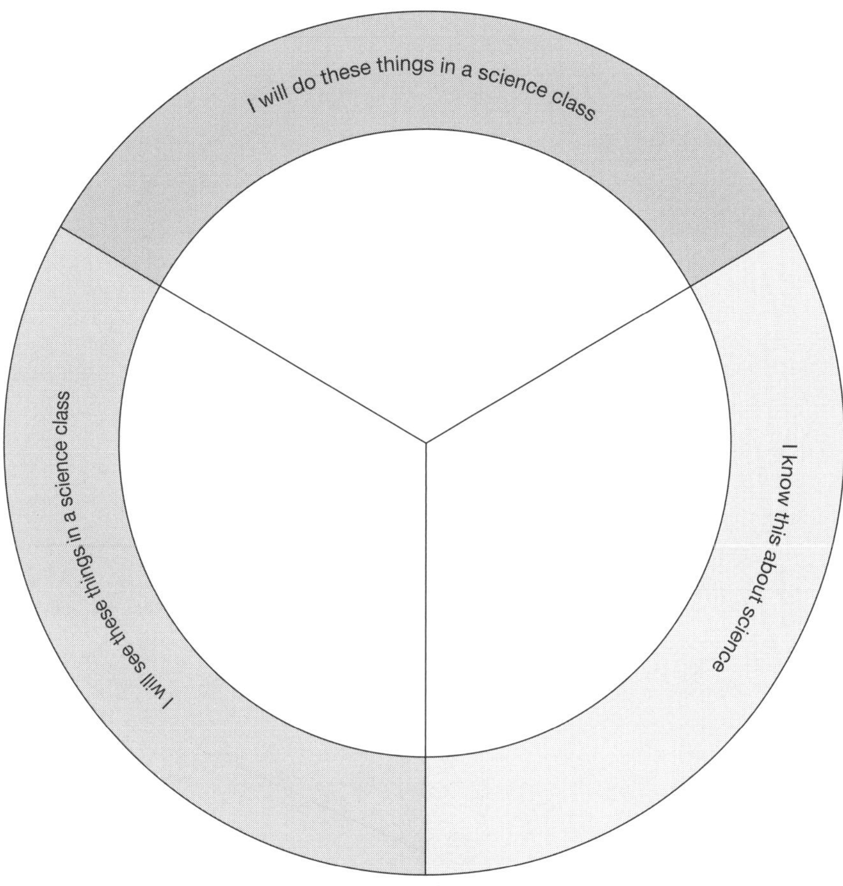

2 List five examples of the use of science in your home or the community.

____

____

____

____

____

____

1.2 The Bunsen burner

Science understanding

FOUNDATION	STANDARD	ADVANCED

1 Use the words in the box below to label parts A to H of the Bunsen burner in Figure 1.2.1.

airhole	barrel	base	cooler part of flame
collar	cone of unburnt gas	gas hose	hottest part of flame

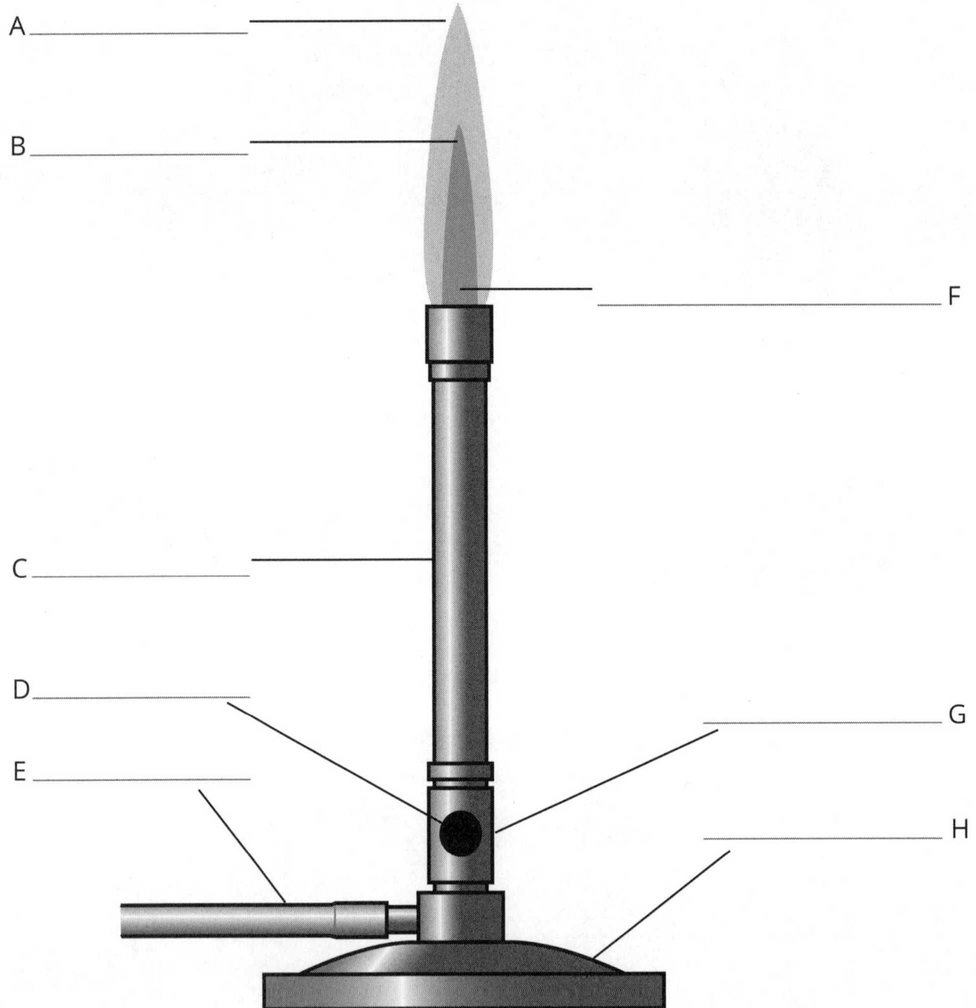

Figure 1.2.1

1.2 The Bunsen burner

(2) Identify which of the diagrams in Figure 1.2.2 correctly show the flames that can be produced by a Bunsen burner. (More than one answer is possible.)

____

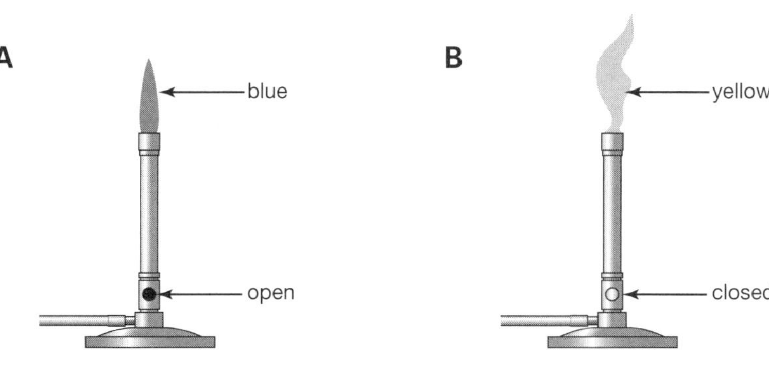

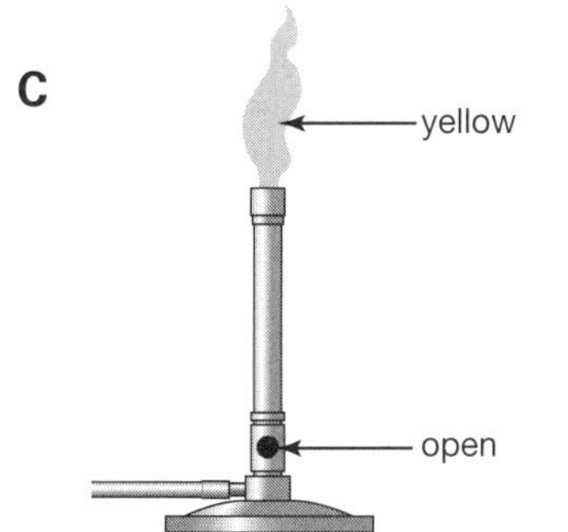

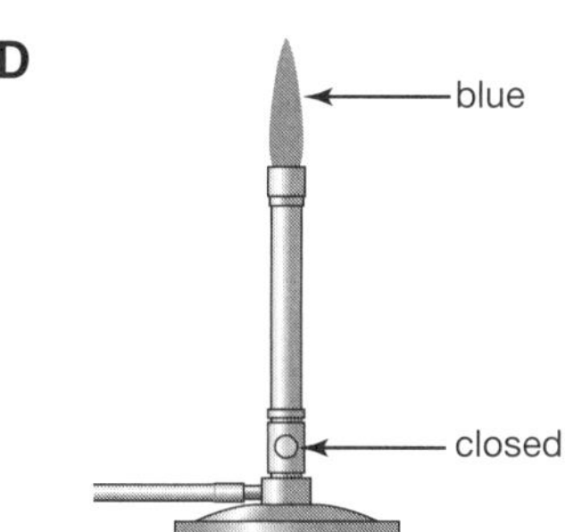

Figure 1.2.2

(3) Below is a set of instructions to light a Bunsen burner and to obtain a hot, blue flame.

However, the instructions are out of order. Use the letters (A, B, C, D, E, F, G, H, I, J, K) to list the instructions in the correct order.

A Strike a match.

B Turn the collar so that the airhole is open.

C Turn on the gas.

D A blue Bunsen burner flame should be seen.

E A yellow Bunsen burner flame should be seen.

F Place the Bunsen burner on the bench mat.

G Place the match just above the top of the barrel of the Bunsen burner.

H Connect the gas hose to the gas tap.

I Make sure that the gas hose is flat, not twisted.

J Turn the collar so that the airhole is closed.

K Place a bench mat on the bench.

The correct order is:

___ then ___ then ___ then ___ then ___ then ___ then ___ then ___ then

___ then ___ then ___

RATE MY UNDERSTANDING
Shade the face that shows your rating

1.3 Identifying laboratory equipment

Science understanding

FOUNDATION	STANDARD	ADVANCED

1 Construct lines to connect the names of the laboratory equipment below with their realistic sketch and with their 2D cross-section diagram.

1	 A conical flask	i 
2	 B tripod and gauze mat	ii 
3	 C watch-glass	iii 
4	 D measuring cylinder	iv 
5	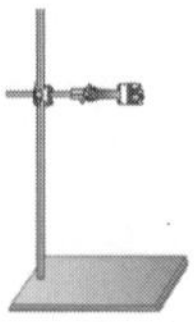 E evaporating dish	v 
6	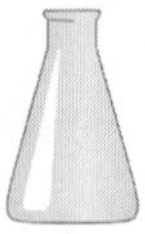 F beaker	vi 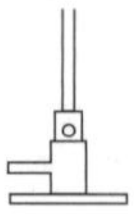
7	 G crucible and lid	vii 
8	 H Bunsen burner	viii 
9	 I test-tube	ix 
10	 J retort stand, bosshead and clamp	x 

2 Summarise your results by writing the number, letter and Roman numeral of each combination, for example 1/C/vi.

1/____/____ 2/____/____ 3/____/____ 4/____/____ 5/____/____

6/____/____ 7/____/____ 8/____/____ 9/____/____ 10/____/____

RATE MY UNDERSTANDING
Shade the face that shows your rating

1.4 Commonsense safety rules

Science inquiry skills

FOUNDATION	STANDARD	ADVANCED

In the cartoons below, some students are acting in a safe and sensible way while others are putting themselves in danger.

1 Identify what each student is doing right or wrong. Underneath each cartoon, write a commonsense rule for the laboratory that deals with the situation.

(a) ______________________

Rule: ______________________

____

(b) ______________________

Rule: ______________________

____

(c) ______________________

Rule: ______________________

____

(d) ______________________

Rule: ______________________

____

(e) ______________________

Rule: ______________________

____

(f) ______________________

Rule: ______________________

____

(g) ______________________

Rule: ______________________

____

(h) ______________________

Rule: ______________________

____

(i) ______________________

Rule: ______________________

____

1.5 Drawing scientifically

Science inquiry skills

FOUNDATION	STANDARD	ADVANCED

Scientific drawing is different from illustrations in other subjects. It is used to show how the equipment is set up. Scientific drawings are simple and only show the main features of each piece of equipment.

Look carefully at Table 1.5.1, which shows examples of illustrations and scientific drawings of the same objects.

Table 1.5.1

Diagram name	Illustration	Scientific diagram	Diagram name	Illustration	Scientific diagram
beaker			filter funnel		
Bunsen burner			test-tube		
conical flask			tripod and gauze mat		

Now look at the list of characteristics of scientific drawings.

1 Highlight the characteristics that are used in scientific drawings. Cross-out the characteristics that are not correct.

Characteristics of a scientific drawing:

- three-dimensional (3D): shows depth, height and width
- open at the top unless the equipment has a lid
- drawn using coloured pencils
- drawn in pen
- two-dimensional (2D): shows height and width
- drawn to be attractive, not accurate
- drawn freehand with some wobbly lines
- drawn in pencil
- drawn to take a full page
- drawn in correct proportions
- drawn using a ruler for straight lines
- large enough to show detail but not to take up a lot of space
- proportions are not important
- provides a general idea of equipment and setup
- provides an accurate image of equipment and setup

② Look at the drawings of the scientific equipment set up for an experiment in Figure 1.5.1. One is an illustration and the other is a correctly drawn scientific drawing.

(a) In the space provided below each drawing, correctly identify and label the illustration and the scientific drawing.

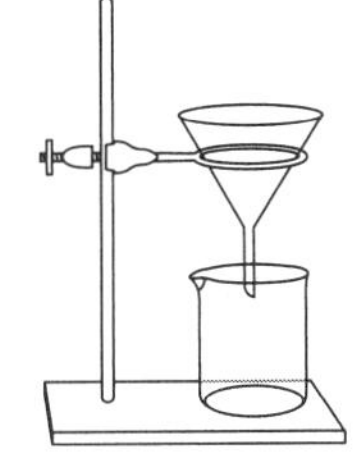

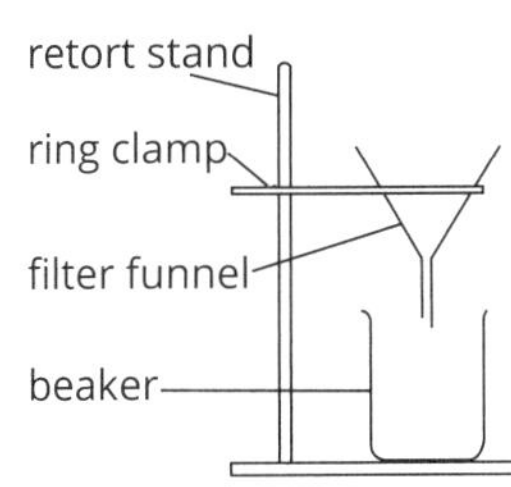

______________________ ______________________

Figure 1.5.1

(b) Compare the scientific drawings in questions 1 and 2(a). You will notice that there are some extra characteristics of scientific drawings. List these features below:

____

____

____

③ In Figure 1.5.2 there are three illustrations of scientific equipment used for experiments. In the spaces provided, draw a labelled scientific drawing of each.

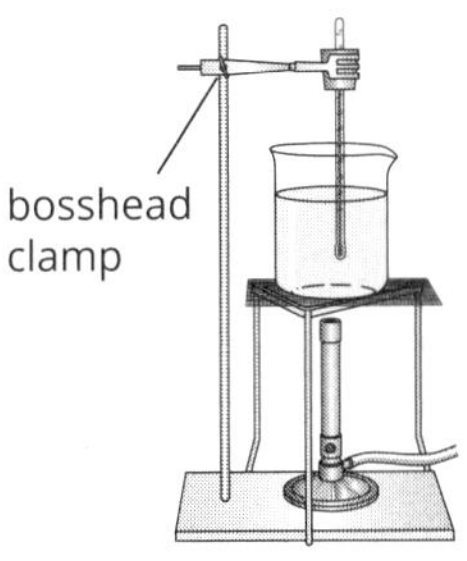

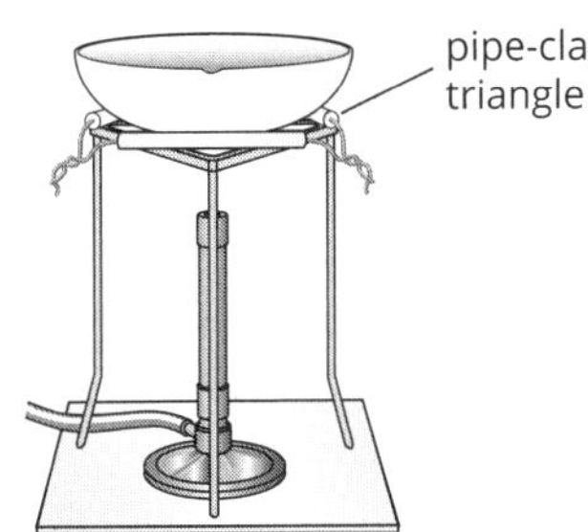

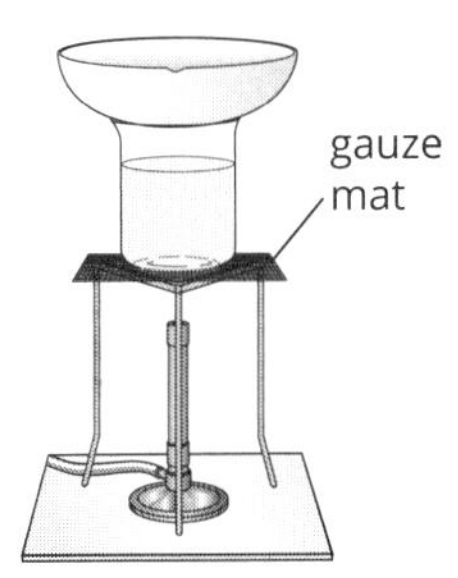

Figure 1.5.2

RATE MY UNDERSTANDING
Shade the face that shows your rating

1.6 Taking measurements

Science inquiry skills

FOUNDATION	STANDARD	ADVANCED

Processing & Analysing

1 State what measurements each of the instruments in Figure 1.6.1 is showing.

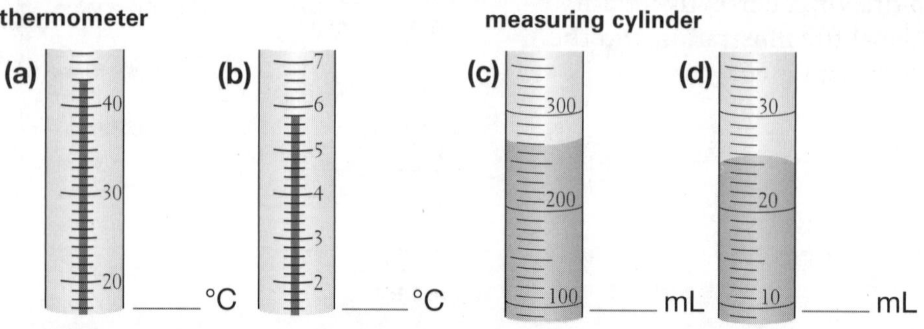

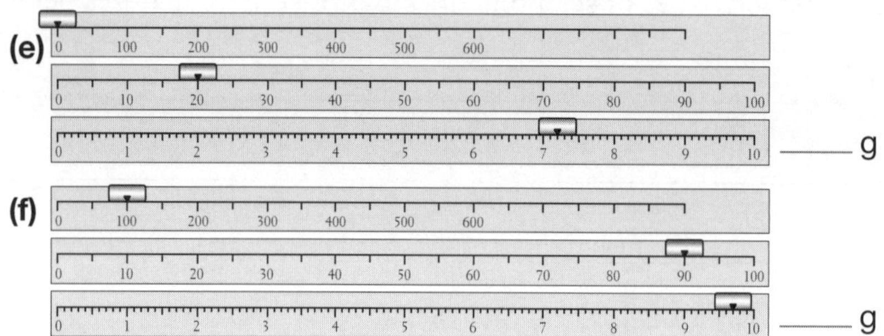

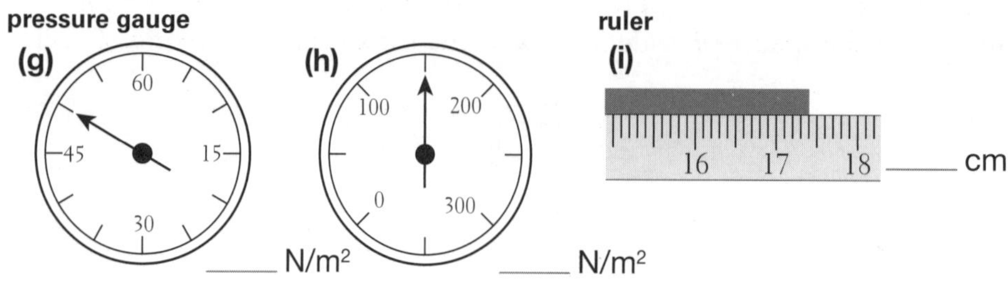

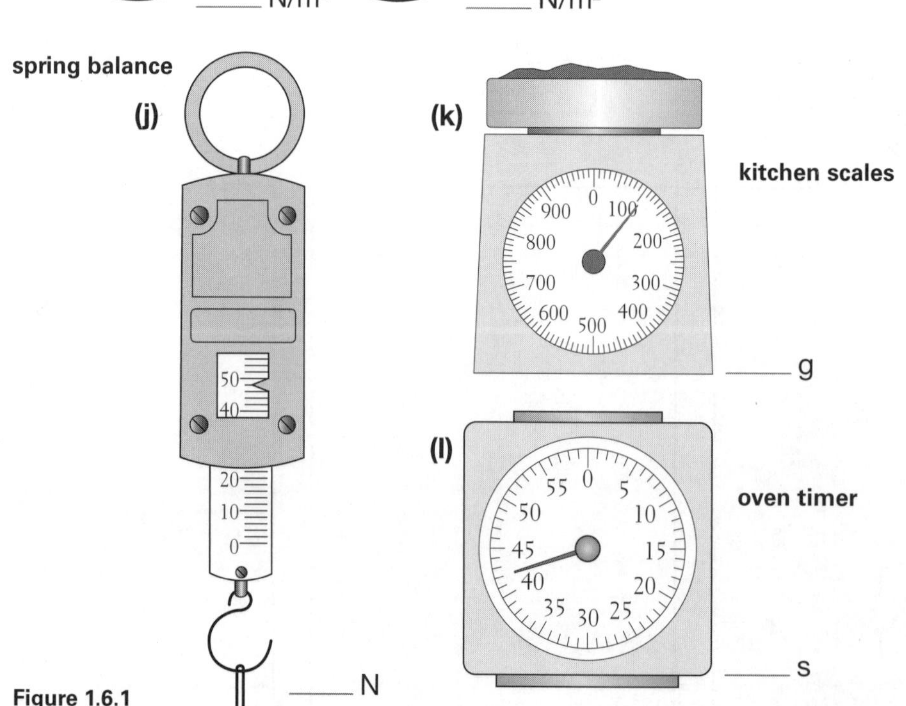

Figure 1.6.1

RATE MY UNDERSTANDING
Shade the face that shows your rating

1.7 Analysing graphs

Science inquiry skills

FOUNDATION	STANDARD	ADVANCED

Processing & Analysing

> **greenhouse gas** *(n)* a gas in the Earth's atmosphere that causes the atmosphere to heat up. This is called the 'greenhouse effect'.
>
> **hydroelectric power station** *(n)* a power station that uses the energy of running water to make electrical energy

This pie graph in Figure 1.7.1 shows the different greenhouse gases in the Earth's atmosphere.

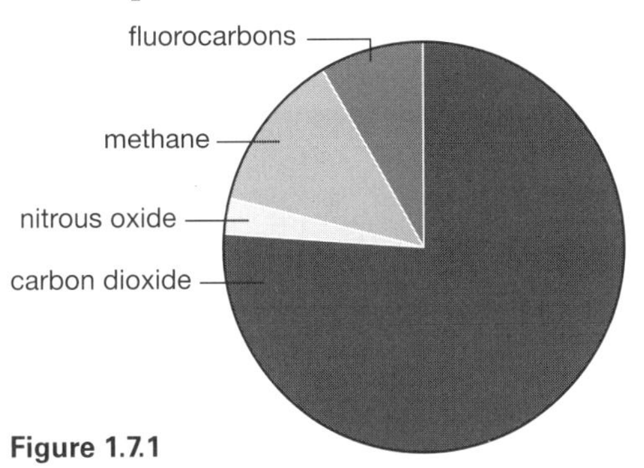

Figure 1.7.1

1 From the list below, identify (circle your answer) what percentage of greenhouse gases is carbon dioxide:

A 10% **B** 25% **C** 50% **D** 75%

2 Methane is a far more powerful greenhouse gas than carbon dioxide. Despite this, most scientists are more worried about carbon dioxide than methane. Use the graph to suggest why.

____

____

3 Use Table 1.7.1 to identify the most likely percentages of greenhouse gases in the atmosphere. Choose A, B, C or D. ________

Table 1.7.1

Percentages of greenhouse gases in the atmosphere					
	Carbon dioxide	Methane	Nitrous oxide	Fluorocarbons	Total
A	50%	25%	13%	12%	100%
B	76%	6%	13%	5%	100%
C	76%	13%	3%	8%	100%
D	61%	13%	13%	13%	100%

4 This bar graph compares the amount of electricity generated by five Australian hydroelectric power plants (measured in megawatts).

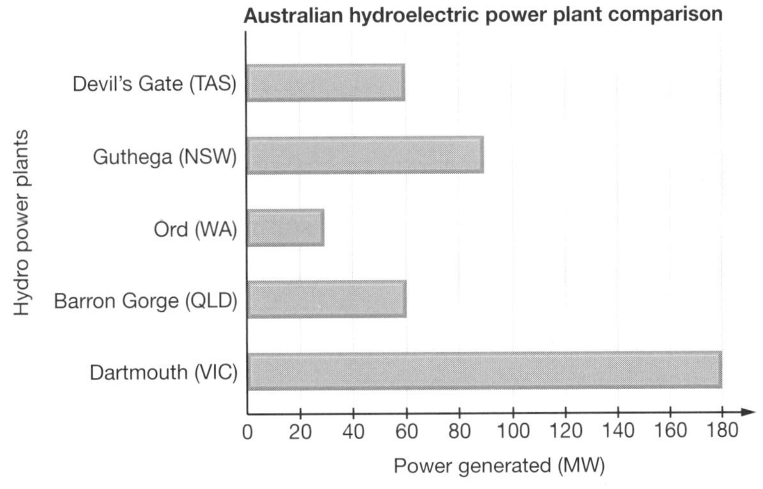

1.7 Analysing graphs

(a) Name the two plants that generate the same amount of power.

______________________ ______________________

(b) Specify how much more power Dartmouth produces than Ord, Barron Gorge and Guthega, by ticking the correct box in the table below.

Table 1.7.2

Dartmouth power generation compared to other power plants			
Amount of power produced	**Ord**	**Barron Gorge**	**Guthega**
about twice as much			
about three times as much			
about four times as much			
about six times as much			

5 The line graph below shows how much energy is used at different times of the day in Melbourne in winter and summer.

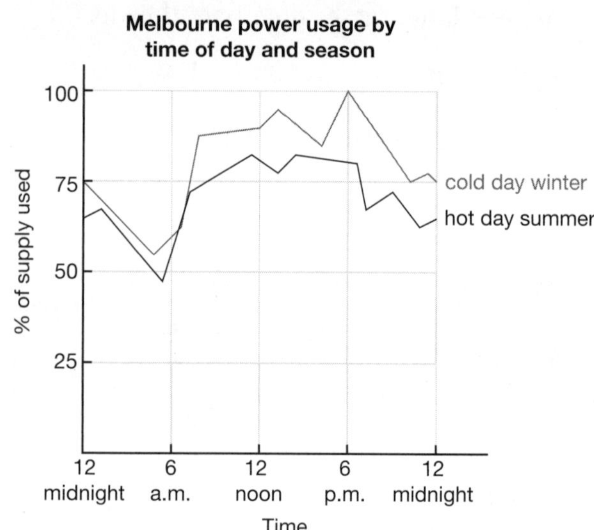

(a) State the maximum percentage of energy supply used in Melbourne. ______________

(b) Specify the time and season when the maximum occurs. ______________

(c) State the minimum percentage of energy supply used. ______________

(d) Specify the time and season when the minimum occurs. ______________

(e) Suggest reasons why these maximums and minimums would occur at those times and in those seasons.

____

____

(f) Compare winter energy consumption with summer energy consumption.

____

____

RATE MY UNDERSTANDING
Shade the face that shows your rating

1.8 Analyse an experiment

Science inquiry skills

FOUNDATION	STANDARD	ADVANCED

Processing & Analysing

As part of her science project, Chloe wanted to find out how big salt crystals would grow under different conditions. She set up two 250 mL beakers and added 100 mL of water and 8 large spatulas of salt to both. She boiled both for 5 minutes on a hotplate. She then placed one beaker in the fridge and the other in a warm, dark place (Figure 1.8.1).

She left both beakers for a day and then sketched what she saw.

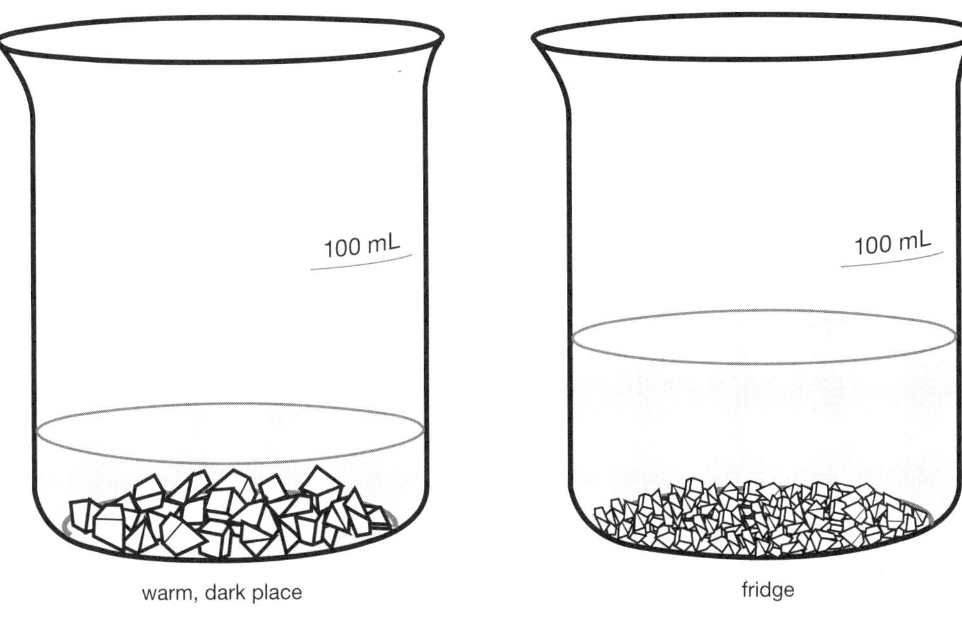

Figure 1.8.1

1. State what the purpose or aim was for Chloe's experiment.

2. Identify the independent variable in Chloe's experiment. This is the variable that Chloe changed in her experiment.

3. Identify the controlled variables in the experiment. These are the variables that Chloe controlled or kept the same.

1.8 Analyse an experiment

4. Explain why Chloe used the same volume of water and the same amount of salt in both beakers.

5. Use the diagrams to list her observations.

6. Identify the best conclusion for Chloe's experiment.

   A The experiment was fun and I learnt a lot.

   B Salt dissolves in water.

   C Salt forms crystals.

   D Cooler temperatures produced smaller crystals.

7. Chloe then wanted to test how the size of crystals depended on the amount of salt added. Construct a dot-point method that she should follow.

RATE MY UNDERSTANDING
Shade the face that shows your rating

1.9 Literacy review

Science understanding

FOUNDATION	STANDARD	ADVANCED

1 Carefully read each definition in the table below. In the space provided, write the correct term for each definition. A list of terms is provided below the table.

Definition	Term
the study of space	
the study of all living things	
the study of chemicals	
the study of the environment	
the study of the Earth	
the study of energy and forces	
the study of behaviour	
the place where scientists work	
poisonous substance	
a special glass that most beakers are made from	
an educated guess	
a factor that may change the result of an experiment	
patterns in results	
the objective or what you are trying to do	
heating device	
amount of liquid measured in litres or millilitres	
curved surface of a liquid	
this is turned to change the type of flame on a Bunsen burner	
matter, measured in grams or kilograms	

Terms: aim, astronomy, biology, chemistry, collar, ecology, geology, hotplate, hypothesis, laboratory, mass, meniscus, physics, psychology, Pyrex, toxic, trends, variable, volume

2 Study the groups of words in the table below. In the spaces provided next to each group:

(a) write down what each group has in common

(b) write one or two sentences that includes at least three of the words. Underline the words used.

Word group	What they have in common	Write a sentence using at least three of the words in the group
(i) beaker spatula test-tube measuring cylinder filter funnel tripod		
(ii) safety glasses follow instructions tie back hair use tongs wear lab coat do not smell chemicals stand test-tubes in rack		
(iii) centimetres millimetres kilograms celsius litres		
(iv) discussion purpose results materials conclusion hypothesis procedure		

RATE MY UNDERSTANDING
Shade the face that shows your rating

1.10 Thinking about my learning

Reflect on what you have you learnt about the subject of science. It may be helpful to look back through your notebook.

Identify six of the most important things you learnt that helped you understand the subject better.

In each of the boxes provided, write down one important thing you learnt. For each point, explain why it is so important to your learning.

I have learnt that . . . I found this important because . . .	I have learnt that . . . I found this important because . . .
I have learnt that . . . I found this important because . . .	I have learnt that . . . I found this important because . . .
I have learnt that . . . I found this important because . . .	I have learnt that . . . I found this important because . . .

CHAPTER 2

Properties of substances

2.1 Knowledge preview

Science understanding

FOUNDATION | STANDARD | ADVANCED

1 Look at the pictures of substances in Figure 2.1.1 below. Use the Venn diagram to decide in which category each picture belongs. Write the name of the substance in the appropriate space.

Figure 2.1.1

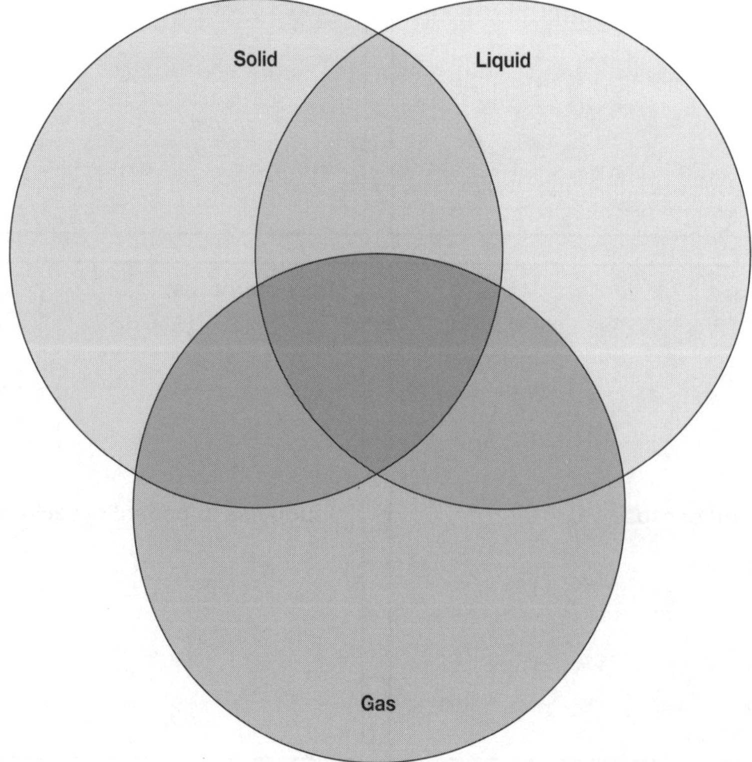

2 Which substances were easy to classify as solid, liquid or gas in the Venn diagram?

____

3 Which were the most difficult to classify? Why?

____

____

____

4 Add 3 more of your own examples to each circle in the Venn diagram.

5 (a) What could be a possible name for a group of substances that fit into more than one of these circles?

____

(b) What are the characteristics of these substances?

____

____

6 Write a descriptive word in each section of the Y-charts below to describe what solids, liquids and gases LOOK LIKE, SOUND LIKE, FEEL LIKE.

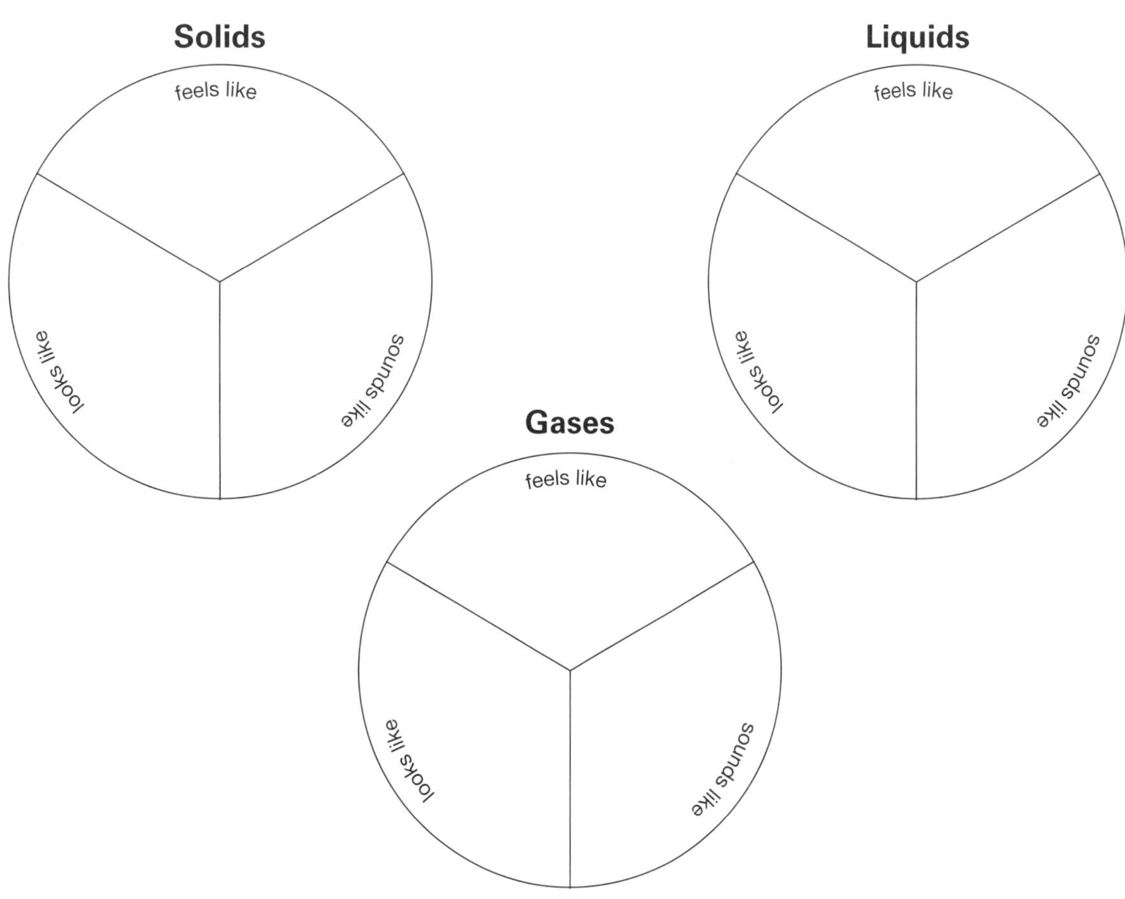

2.1 Knowledge preview

7 Look at the key word list below. Make a list of the words you know and write a definition or description, or give an example of what it is.

boiling	chemical properties	compressed	condensation
density	evaporation	freezing	mass
melting	particle model	physical properties	sublimation
states of matter	steam	volume	

2.2 Biodegradability

Science inquiry skills

FOUNDATION	STANDARD	ADVANCED

1 Define the term biodegradable.

2 List signs that indicate a substance is biodegradable.

3 Classify whether the following substances and objects are biodegradable or not by placing a tick in the correct column.

Substance or object	Biodegradable	Non-biodegradable	Substance or object	Biodegradable	Non-biodegradable
autumn leaves			fruit salad		
pebbles			glass bottle		
polystyrene cup			woollen jumper		
plastic fork			wooden log		
dead rat			lamb chop		

4 Look carefully at John's lunch box and its contents (Figure 2.2.1).

(a) Use a green marker to highlight all of the substances in John's lunch box that are biodegradable.

(b) Use a yellow marker to highlight all of the substances that are non-biodegradable.

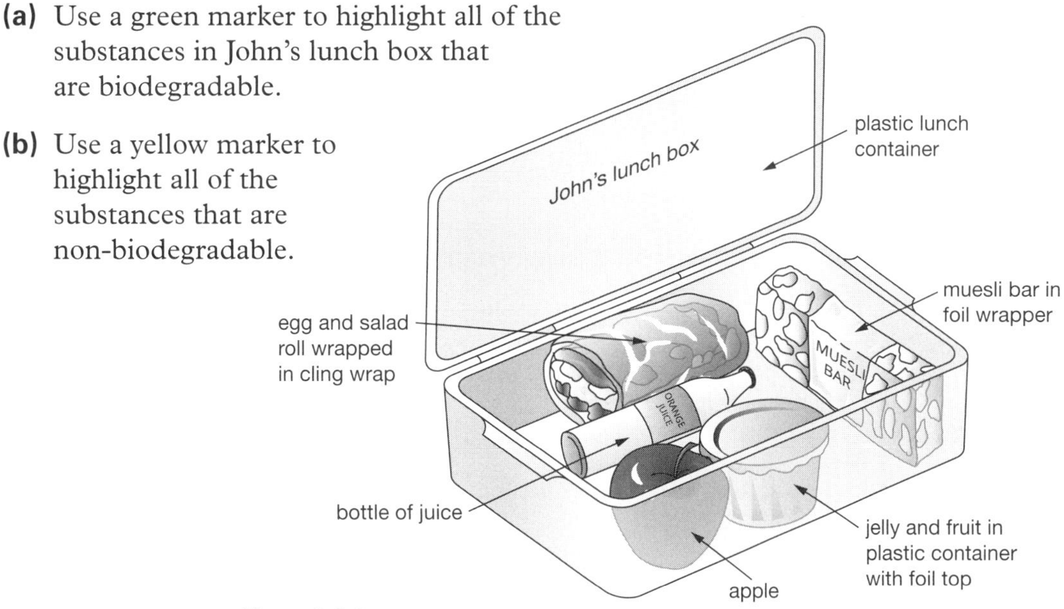

Figure 2.2.1

5 Explain why we should all recycle non-biodegradable substances.

RATE MY UNDERSTANDING Shade the face that shows your rating			☹

2.3 The particle model

Science inquiry skills

FOUNDATION	STANDARD	ADVANCED

Processing & Analysing

Four balloons were blown up to different sizes in different rooms of a house. The temperature of each room was different. The balloons are shown in Figure 2.3.1.

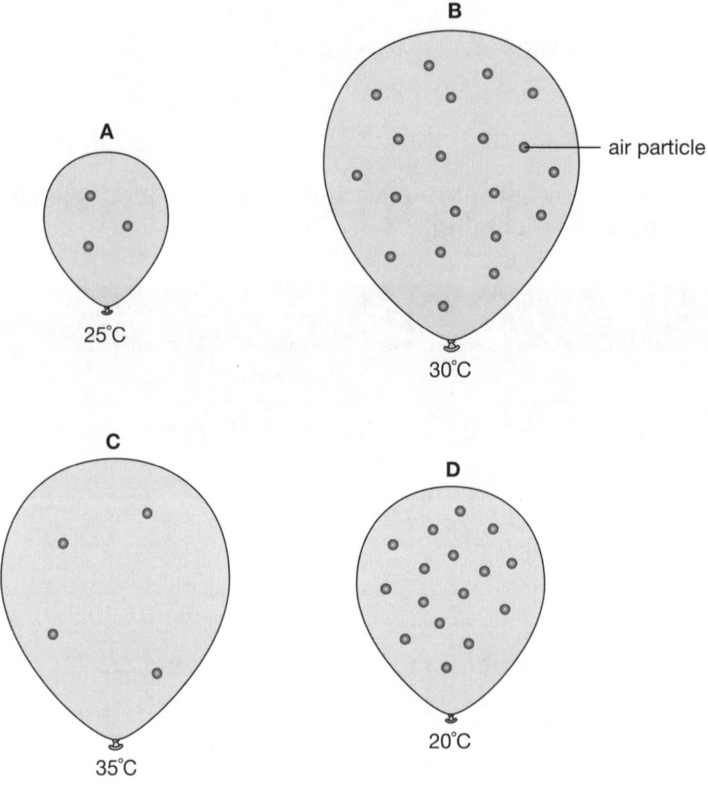

Figure 2.3.1

1. Identify the balloon (A, B, C or D):

   (a) in which the air particles would be moving the fastest ________

   (b) in which the air particles would be moving the slowest ________

   (c) in which the air particles are furthest apart ________

   (d) in which the air particles are closest to each other ________

   (e) that would be the heaviest ________

   (f) that would be the lightest ________

   (g) that has the most space/greatest volume ________

   (h) that has the least space/smallest volume ________

   (i) with the densest air ________

   (j) with the least dense air ________

RATE MY UNDERSTANDING
Shade the face that shows your rating   

2.4 Changes of state

Science understanding

FOUNDATION | **STANDARD** | ADVANCED

The three main states of matter—solid, liquid and gas are shown in Figure 2.4.1.

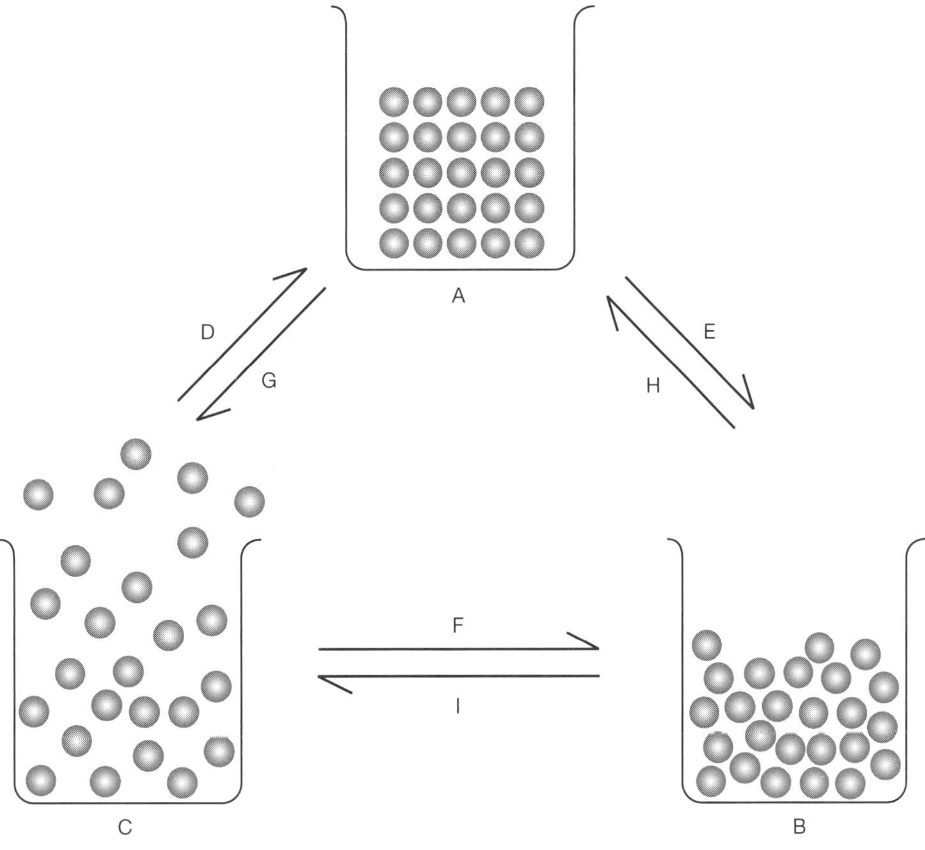

Figure 2.4.1

1 Identify which diagram (A, B or C) best represents a:

(a) solid _______ **(b)** liquid _______ **(c)** gas _______

2 Identify which arrows (D, E, F, G, H or I) represent the following changes of state:

(a) melting _______

(b) freezing _______

(c) evaporation _______

(d) condensation _______

(e) sublimation _______

(f) deposition _______

(3) Explain why the melting point of a substance is the same as its freezing point.

____

____

2.4 Changes of state

Naphthalene is a smelly chemical commonly used in mothballs. Some flakes of naphthalene were heated up until they melted then boiled. The graph below shows the important stages in this heating.

mothballs *(n)* small balls of chemicals, usually naphthalene, used to protect clothing from moths

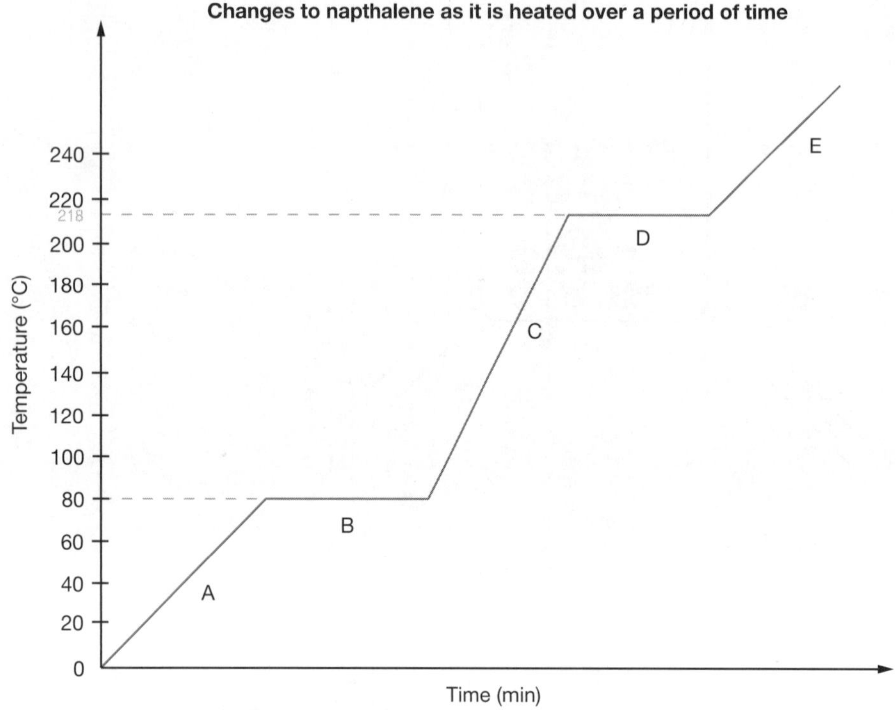

4 Identify which section of the graph (A, B, C, D or E) best represents when naphthalene was:

(a) all gas ________________

(b) in both gaseous and liquid states ________________

(c) all liquid ________________

(d) in both liquid and solid states ________________

(e) all solid ________________

5 Use the graph to predict the melting point of naphthalene. Which of the following options is correct?

A 0°C **C** 100°C

B 80°C **D** 218°C ________________

6 Use the graph to predict the boiling point of naphthalene. Which of the following options is correct?

A 0°C **C** 100°C

B 80°C **D** 218°C ________________

RATE MY UNDERSTANDING
Shade the face that shows your rating

2.5 Cooling curve

Science understanding

FOUNDATION | **STANDARD** | ADVANCED

Salty water was being heated on a hotplate. When it boiled it was then removed from the hotplate and placed in a freezer to cool. Its temperature was measured every minute. The measurements taken are shown in Table 2.5.1 below.

Table 2.5.1

Time (minutes)	0	1	2	3	4	5	6	7	8	9	10	11	12	13	14	15	16
Temperature (°C)	104	103	102	91	82	70	59	51	40	28	20	11	2	–1	–4	–4	–4

1. **(a)** Construct a graph by plotting these values on the grid provided.

   (b) Write a title for the graph.

2. Read each statement and indicate whether it is true or false by circling the correct answer.

   (a) After 2 minutes in the freezer, the salty water had a temperature of 104°C. True / False

   (b) The temperature reached 40°C after 8 minutes. True / False

   (c) The temperature dropped by 12°C between the 8th and 9th minute. True / False

   (d) Every minute the temperature dropped by 12°C. True / False

   (e) The data was collected over 15 minutes. True / False

   (f) Over the time data was collected, the temperature dropped by 108°C in total. True / False

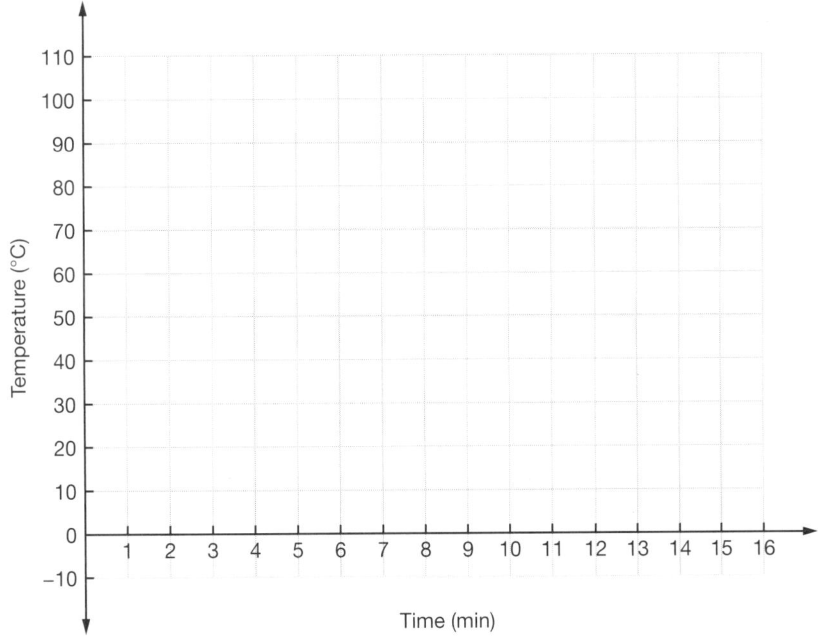

3. Use your graph to estimate:

   (a) the boiling point of salty water ______________________________

   (b) the freezing point of salty water. ______________________________

2.6 Incomplete experiment report

Science understanding

FOUNDATION | STANDARD | ADVANCED

A cooling curve experiment was conducted to obtain the data shown in Worksheet 2.5, page 23. A complete investigation report was not provided.

Use the information in Worksheet 2.5 and the practical investigation report checklist in Figure 2.6.1 below to complete the following tasks.

controlled variable *(n)* factor that stays the same

dependent variable *(n)* what is being measure or tested

independent variable *(n)* what is being changed and how it is being changed to test the dependent variable

Practical investigation report checklist

Title

- □ what was investigated

Purpose

- □ the purpose describes what you wanted to show, prove or find out in an investigation
- □ can be a statement or a question
- □ one or two sentences
- □ often written as 'To investigate the effect of ... on ...'

Hypothesis

- □ a hypothesis is a prediction about the result of your investigation
- □ a short statement
- □ describes the different things you tested (these are called dependent and independent variables)
- □ not always included in a scientific report

Materials

- □ a list of all the important equipment, chemicals and materials that you used
- □ includes quantities of substances and sizes of equipment

Procedure

- □ the procedure or method is a detailed list of what you did in the experiment, in the exact order you did it
- □ written in short, numbered steps
- □ includes the quantities you used (e.g. 5 g, 2 spatula loads, 10 mL)
- □ can include diagrams of the experiment (2D scientific diagrams)

Results

- □ results are a record of all the observations and measurements you took during the investigation
- □ observations can be written and can include diagrams, photos and videos
- □ written observations are best presented in a table
- □ include any graphs or calculations

Review

- □ an analysis of your observations and measurements
- □ analyse any table, spreadsheet or graph you produced
- □ compare your findings with other groups or with information found from textbooks or the internet
- □ evaluate how you could make your investigation better
- □ construct a short conclusion that summarises what you found out in the experiment
- □ use your conclusion to evaluate how accurate your hypothesis was

Figure 2.6.1

(1) On the checklist, tick all the parts of a report that are covered in Worksheet 2.5.

2 What do you think the aim of the experiment was?

____

____

(3) Name the dependent variable.

____

____

4 Determine whether the data collected is qualitative or quantitative.

qualitative data *(n)* data collected as descriptions, e.g. hot

quantitative data *(n)* data collected as numbers, e.g. 35°C

5 List the ways data is presented.

(6) A materials list is not included in Worksheet 2.5. Write the materials list exactly as it would appear on a practical investigation report.

(7) Worksheet 2.5 does not include a description of the method used for the experiment, as it should appear on a practical investigation report. Write the method section of the practical investigation.

8 This experiment was only carried out once. What could be done to ensure the results are fair?

RATE MY UNDERSTANDING
Shade the face that shows your rating  

2.7 Archimedes

Science as a human endeavour

FOUNDATION	STANDARD	**ADVANCED**

density *(n)* the mass of material that is packed into an object

mass *(n)* the amount of material in an object

prism *(n)* a solid object such as a cube or cylinder

volume *(n)* the amount of space that an object occupies

Archimedes lived from about 287 to 212 BCE. He was born in Syracuse, on the island of Sicily. Although it is now a part of Italy, Syracuse was then a colony of ancient Greece. Little is known about Archimedes' life and most of what we do know comes from stories written by Roman historians long after his death.

Archimedes and density

According to a Roman story, illustrated in Figure 2.7.1, Archimedes worked out how to calculate the density of an irregular object. Density is the mass of an object divided by its volume. Hiero II, the king of Syracuse, suspected that his goldsmith had cheated him by substituting cheaper silver for gold in a wreath the king had ordered to be made.

Figure 2.7.1 Archimedes found a non-destructive way of finding the volume of an object.

Archimedes was asked to work out whether the wreath was pure gold or not. He knew that if the wreath contained silver, then its density would be less than that of gold. In order to work out the wreath's density he needed to measure both the mass and volume of the wreath. Mass could be easily measured using scales, but he wondered how he could measure the volume of such an irregularly shaped wreath. One way was to melt down the wreath, make it into a regular box-shaped prism, and then calculate its volume. However, this would have destroyed the wreath. Archimedes needed to find a non-destructive way of testing the wreath.

While pondering this question, Archimedes supposedly took a bath. On lowering himself in, he noticed that the water level rose. He instantly realised that the water rose by the same volume as his body. He realised he could use the same method to measure the volume of the wreath! Excited by his discovery, Archimedes allegedly ran naked into the streets shouting 'Eureka, eureka!'

1 Propose reasons why the wreath could not be melted down.

____

____

2 Propose what a destructive test of the wreath would be.

____

____

3. Destructive tests would never be carried out in the following situations. For each situation, propose a reason why.

(a) Testing the strength of the Sydney Harbour Bridge.

(b) Testing the amount of chemical pollutants that would kill people.

(c) Testing the force in a punch that would cause brain injury.

4. Assume that Archimedes' experiment had the following results:

- the wreath's mass was 80 g
- the volume of displaced water was 5 cm^3

Calculate the density of the wreath.

5. Do you think that Archimedes' experiment provided him with all the information he needed to prove whether the wreath was made of pure gold or not? Justify your answer.

6. Explain what further test/tests need to be done to find whether the wreath is or is not made of pure gold.

RATE MY UNDERSTANDING
Shade the face that shows your rating

2.8 Density calculations

Science understanding

FOUNDATION | **STANDARD** | ADVANCED

If an object has a regular shape like a cube or a box, then you don't need to use a measuring cylinder to find its volume. You can use maths instead. The volume of a box can be calculated using the formula:

Volume = length × width × height

$V = lwh$

(1) **(a)** Use the formula $V = lwh$ to calculate the volume of the rectangular prisms shown in Figure 2.8.1.

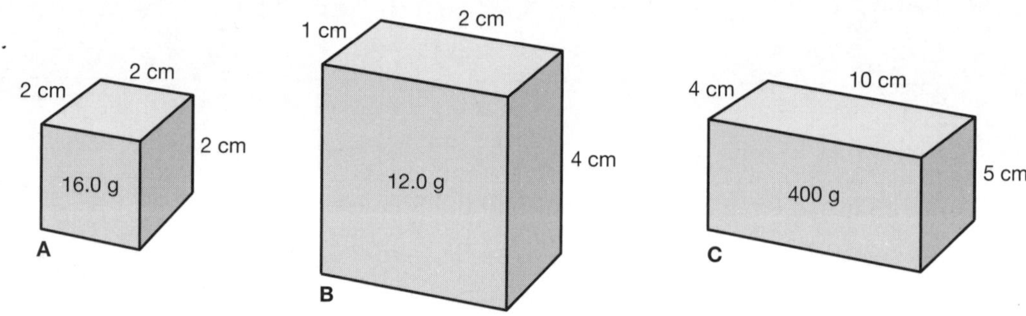

Figure 2.8.1

Prism A: $V =$ ______________________ cm³

Prism B: $V =$ ______________________ cm³

Prism C: $V =$ ______________________ cm³

(b) Use the masses given for each of the prisms to calculate their densities.

Prism A: $d = \frac{m}{V} =$ __________ / __________ = __________ g/cm³

Prism B: $d = \frac{m}{V} =$ __________ / __________ = __________ g/cm³

Prism C: $d = \frac{m}{V} =$ __________ / __________ = __________ g/cm³

② **(a)** Calculate the volume of the irregular object shown in Figure 2.8.2 that has been put inside a measuring cylinder containing some water.

V = ________________ mL = ________________ cm³

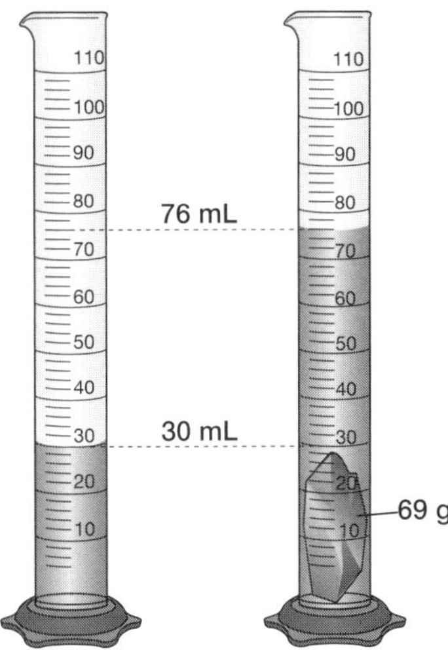

Figure 2.8.2

(b) Use the mass given for the irregular object in part (a) to calculate its density.

Irregular shape: $d = \frac{m}{V}$ = ______________ / ______________ = ______________ g/cm³

③ The mass of an unknown liquid was determined by the method shown.

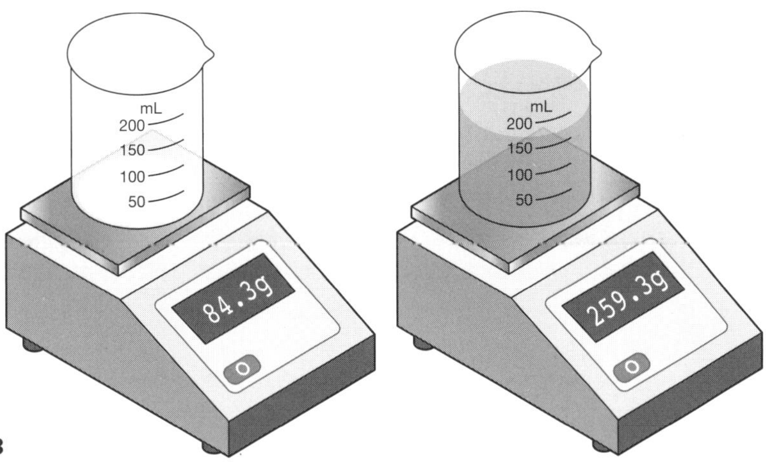

Figure 2.8.3

(a) Use the information in Figure 2.8.3 to calculate the density of the liquid.

____

(b) From its density, propose what the unknown liquid is most likely to be.

____

④ Prism A in question 1 was dropped into the measuring cylinder in Figure 2.8.4. Modify the second measuring cylinder by marking the level the water should rise to.

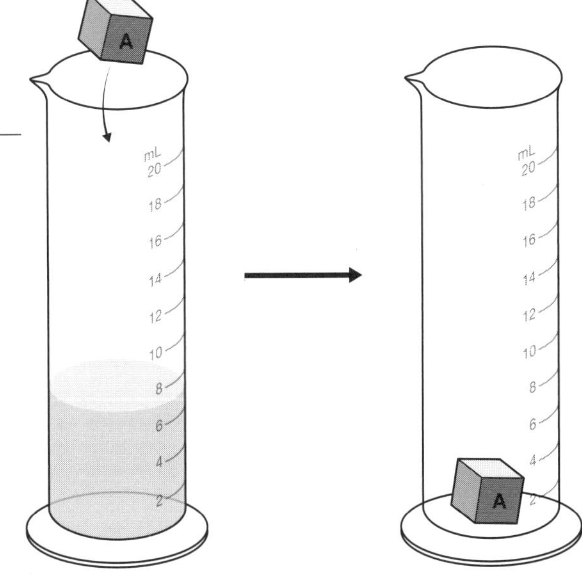

Figure 2.8.4

RATE MY UNDERSTANDING
Shade the face that shows your rating

2.9 Literacy review

Science understanding

FOUNDATION	STANDARD	ADVANCED

1 Look at the list of properties in the box below.

50 millilitres of water in a glass	mixing sand in water	burning a timber log
iceberg in Antarctica	rusting iron fence	dynamite explosion

(a) Identify the physical properties and highlight them in yellow.

(b) Identify the chemical properties and highlight them in green.

(c) Explain the difference between physical and chemical properties.

____

____

2 Think about solids, liquids and gases by referring to water particles. Use the spaces below to:

(a) describe/draw what water particles look like in solid (ice), liquid (water) and gas (water vapour) states

(b) describe what happens if the temperature changes, as prompted on the table.

Solid (ice)	Liquid (water)	Gas (water vapour)
Description/drawing of particles	Description/drawing of particles	Description/drawing of particles
If temperature increases	If temperature increases If temperature decreases	If temperature decreases

HINT

Boiling point: at 100°C water turns into vapour

RATE MY UNDERSTANDING
Shade the face that shows your rating

2.10 Thinking about my learning

Tick the square that best matches your understanding for each of the big ideas.

	Big ideas	I still need help with this	I understand this	I understand this well and can teach someone about this
Science understanding	I can draw a model to represent a solid, liquid and a gas.			
	I can give examples of solids, liquids and gases.			
	I understand why we use the particle model to represent solids, liquids and gases.			
	I can explain how the particles in solids, liquids and gases react when there is a change in temperature.			
	I know what physical and chemical properties are and can give examples of each.			
	I can use chemical and physical properties to describe solids, liquids and gases.			
	I can use the correct words to describe the processes of how solids, liquids and gases change from one state to another.			
	I can explain the difference between mass, volume and density.			
	I can give examples of substances that are more dense and less dense than water.			
Science inquiry skills	I can follow a method and set up experiments involving solids, liquids and gases.			
	I can write experiment reports and discuss and explain my experiment results.			
	I can identify problems with experiment methods and results and suggest improvements that could be made.			
	I can work safely in the science laboratory.			
Science as a human endeavour	I know and can discuss the issues relating to biodegradable and non-biodegradable substances and the impact they have on the environment.			
	I can explain how our knowledge of particles and the particle model has developed over time.			

CHAPTER 3

Earth resources

3.1 Knowledge preview

Science understanding

FOUNDATION	STANDARD	ADVANCED

Work in a small group to assess your knowledge on the topic of renewable and non-renewable resources.

1 What is a definition of a natural resource?

____

____

____

2 List at least ten natural resources.

____

____

____

____

____

3 Coal is a non-renewable resource while air is a renewable resource.

(a) What is a non-renewable resource?

____

____

____

(b) What is a renewable resource?

____

____

____

(c) What is one similarity between coal and air?

____

____

3.1 Knowledge preview

A ______________ B ______________ C ______________ D ______________

Figure 3.1.1

4 Identify the types of energy generation shown in Figure 3.1.1 by labeling each photograph.

5 The following are common terms that are used when discussing energy resources: greenhouse gasses, renewable energy, fossil fuels, solar energy, non-renewable energy. Write a short paragraph using all these terms to demonstrate your knowledge and understanding of energy resources.

____

____

____

____

____

____

6 Water is an extremely important natural resource.

(a) List five ways that water is used.

____

____

____

____

____

(b) Describe two ways that water is managed.

____

____

3.2 Differences in soils

Science understanding

FOUNDATION | **STANDARD** | ADVANCED

decaying (*v*) breaking down, rotting
extract (*v*) to take something out; to remove
organism (*n*) a living thing
ploughing (*v*) using a tool or a machine to turn over the top layer of soil before planting seeds

Soil is a vital resource. Soils differ in many ways, the most important being how they affect plant growth. The three main types of soil are clay, sand and silt. Other types of soils are combinations of these three types. For example, loam has similar proportions of sand, silt and clay. The different types of soil are shown in Figure 3.2.1.

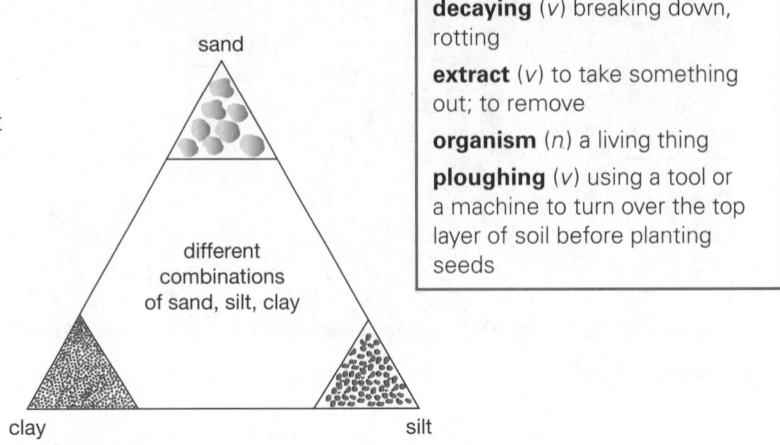

Figure 3.2.1 Types of soil

1 Name the three main types of soil. ______________ ______________ ______________

2 What soil particles would you expect to find in:

sandy clay? ______________ loam? ______________

The sizes of different particles of soil are shown in Figure 3.2.2. Particles less than 0.002 mm are invisible without magnification.

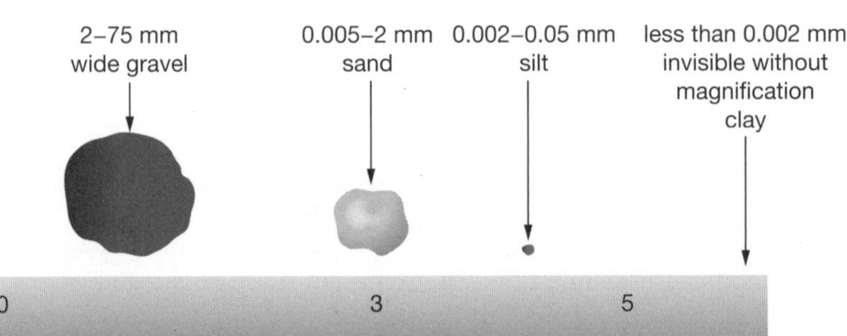

Figure 3.2.2 Relative size of soil particles

3 What is the size of a particle of clay? ______________

4 Compare the size of a particle of silt to a particle of sand. ______________________

The main differences in soils that affect their suitability for plant growth are shown in Table 3.2.1 and Figure 3.2.3.

Table 3.2.1 Features of soils and how they affect plant growth

Soil feature	Description of the feature	How the feature affects plant growth
texture	Texture the size of the particles that make up soil. The main particles in soil are classified as clay, silt or sand. Most soils have various combinations of these particles.	Fine textured soils, such as clay, have tiny pores. These soils have a greater water-holding capacity, restricted aeration and slower decomposition of organic matter to enrich them. Sandy soils have course particles and larger pores. These soils are well drained, have good aeration, allowing growth of plant roots, and have a greater oxygen content which speeds up the decomposition of organic matter.
structure	Structure describes how well the soil particles join up to form lumps, called crumbs. Crumbs have spaces between them, called pore spaces. The crumbs are about 3 mm to 5 mm wide. Soil texture and structure of loam soil is shown in Figure 3.2.3.	Pore spaces let water and air enter the soil to improve plant growth.

3.2 Differences in soils

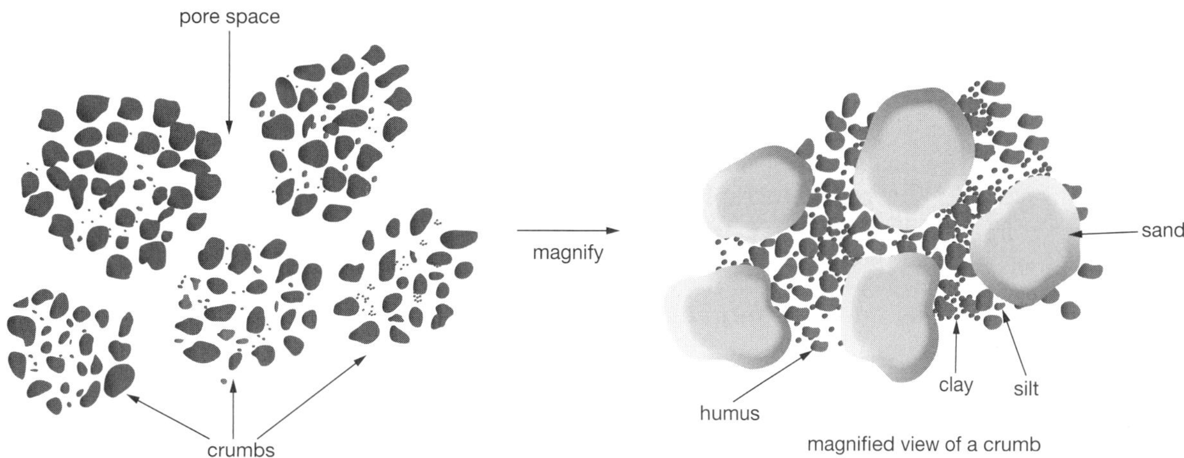

Figure 3.2.3 The structure of a loam soil showing crumbs held together by clay, humus (organic matter), silt and sand.

5 Describe what 'texture' refers to in a soil.

____

6 Define 'pore spaces' in soil.

____

Table 3.2.2 Features of soils and how they affect plant growth

Soil feature	Description of the feature	How the feature affects plant growth
consistency	Consistency is the tendency of soil particles to stick together. Clay soils have a high consistency.	Too much working of soils (such as farmers ploughing) can compact it into a layer like concrete, which is poor for plant growth.
humus content	Humus is decaying organic matter—dead organisms and their wastes.	Plants grow very well in soils high in humus because it holds water and minerals well and helps crumbs to form.
mineral content	Mineral content is the minerals present in the soil.	Minerals are needed for living organisms to help their bodies function and grow.

7 Explain how humus can help soil structure.

____

____

8 Explain how the mineral content of a soil can be important for plant growth.

____

____

9 **(a)** Explain why pore space is important to plants.

____

(b) Propose reasons why farmers need to be careful not to use machinery too much on their soils.

____

RATE MY UNDERSTANDING Shade the face that shows your rating	☺		

3.3 Looking after soil

Science inquiry skills

FOUNDATION	STANDARD	ADVANCED

Processing & Analysing

slope (*n*) not flat; on an angle

To many people who live in cities, soil may not seem to be an important resource. However, much of our food comes from plants grown in the soil and from animals fed on those plants. Protecting soil is an important task that farmers take very seriously.

Soil erosion is a problem in many farming areas. Running water and wind are the two main agents of erosion on farms in Australia. Soil erosion is worse where the land slopes.

Two students, Siobhan and Ivan, wanted to find out how the speed of a stream affects the size of the weathered rock particles it carries. They set up a piece of roof gutter on their back lawn and filled it with a mixture of clay, sand and gravel soil. The gutter was supported so that it was on a gentle slope. A tray at the lower end of the gutter was set up to catch soil that washed away. A hose was set up at the high end of the gutter and the tap was turned on so the water ran fairly slowly into the soil. Figure 3.3.1 shows the set up.

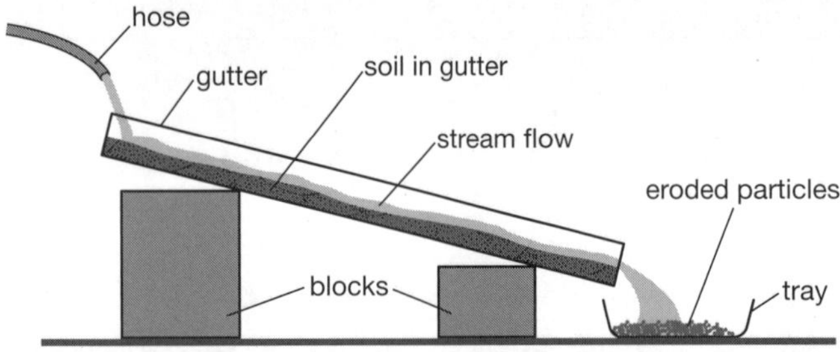

Figure 3.3.1 Experiment set-up

After five minutes they turned off the tap. They collected the soil that was washed into the tray, dried it in the sun, and then weighed it. A microscope was used to measure the size of the largest particles that collected in the tray. They then repeated the experiment with different slopes, making sure they replaced the soil in the gutter each time and used the same number of turns of the tap.

Their results are shown in Table 3.3.1.

Table 3.3.1 Siobhan and Ivan's soil erosion experiment results

Steepness of slope	Stream speed	Mass of particles (g)	Largest particle size (mm)
gentle	slow	18	0.2
moderate	moderate	39	2.0
steep	fast	57	3.0

1 Identify the independent variable (the variable that the students changed) in this experiment.

____

2 Identify the dependent variable (the variable that was being measured) in this experiment.

____

3 Identify five variables the students kept the same in each experiment.

____

____

4 Assess the effect the speed of the stream had on the mass of particles it carried.

____

____

5 Assess the effect the speed of the stream had on the size of the particles it carried.

____

____

6 Explain why a slow stream cannot carry large particles.

____

____

7 **(a)** Look at Figure 3.3.2. Where, in a river's journey from the mountains to the sea, would most of the heavy particles (like gravel) be deposited? Mark this location on the diagram with an X. Explain why heavy particles would be deposited here.

____

____

____

(b) Where in the river's journey would fine particles (like clay) be deposited? Mark the location with a Y. Explain why they would be deposited here.

____

____

____

River carries away sediments as it erodes the mountains

River slope as it enters the sea

Ocean

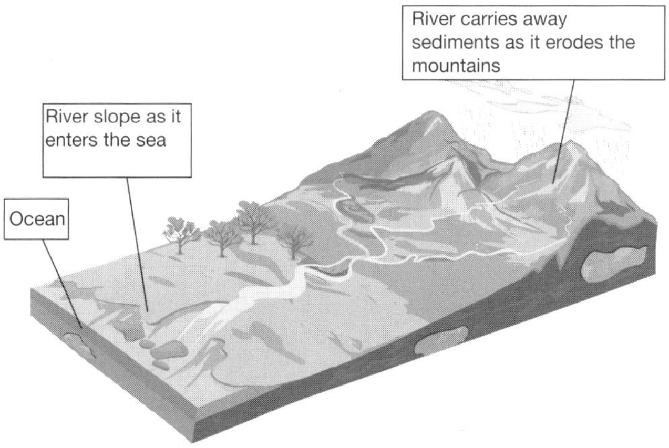

Figure 3.3.2 A river's course from mountains to the sea

3.4 Biofuels inquiry task

Science inquiry skills

FOUNDATION	**STANDARD**	ADVANCED

Processing & Analysing

conserve (*v*) to protect the environment from being destroyed or harmed

fossil fuels (*n*) natural (organic) fuels formed from the remains of former life, e.g. natural gas, coal, petroleum

non-renewable energy (*n*) energy from sources that will not be replenished, e.g. fossil fuels (natural gas, coal, petroleum)

renewable energy (*n*) energy from sources such as the sun (solar) or wind that is not depleted when used

sustainable (*adj*) to use techniques or methods so as to conserve natural resources, and to not use up or deplete them

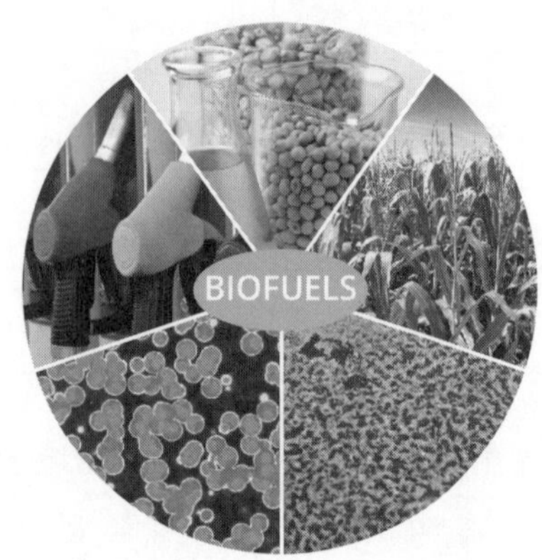

Biofuel is liquid fuel made from living things like plants or algae. Biofuels have the potential to provide greener fuels for the future and are a sustainable and clean alternative to using non-renewable fuels like nuclear fuels, coal, oil and natural gas.

1 Research and list examples of biofuels.

____

____

____

(2) Investigate and discuss the positives, negatives and interesting facts you have discovered about the use of biofuels and complete the PMI chart below.

Plus	Minus	Interesting

3 Choose a biofuel to research and complete a flow chart (like the one below) that demonstrates the source of the biofuel, how the biofuel is processed and what the biofuel is used for. Provide an explanation of each stage in the process. You may use pictures to help you.

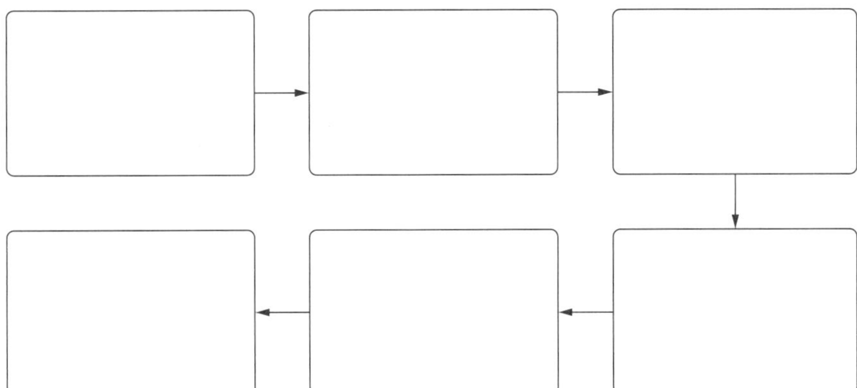

4 Outline and discuss any challenges in the growth, production and use of biofuels.

5 Why do you think that some people believe governments should allocate more time and money into biofuel research, production and use?

RATE MY UNDERSTANDING
Shade the face that shows your rating  

3.5 Water cycle terms

Science understanding

FOUNDATION | **STANDARD** | ADVANCED

1 Draw a line to match each term below with its correct definition.

Term	Definition
evaporation	any water falling out of the sky
percolation	change of state from liquid to water vapour
condensation	evaporation of water from plants
precipitation	the process of water soaking into the soil
transpiration	change of state from water vapour to liquid

(2) In the table below, propose the effect that each change (a), (b), (c) and (d) would have on the rate of evaporation of water. Justify your response in each case.

Change	Effect on evaporation rate	Justification
(a) increased temperature		
(b) increased humidity		
(c) increased air movement		
(d) size of the body of water		

RATE MY UNDERSTANDING
Shade the face that shows your rating

3.6 The water cycle

Science understanding

FOUNDATION | **STANDARD** | ADVANCED

1 Name the process that is occurring at each point of the water cycle in Figure 3.6.1.

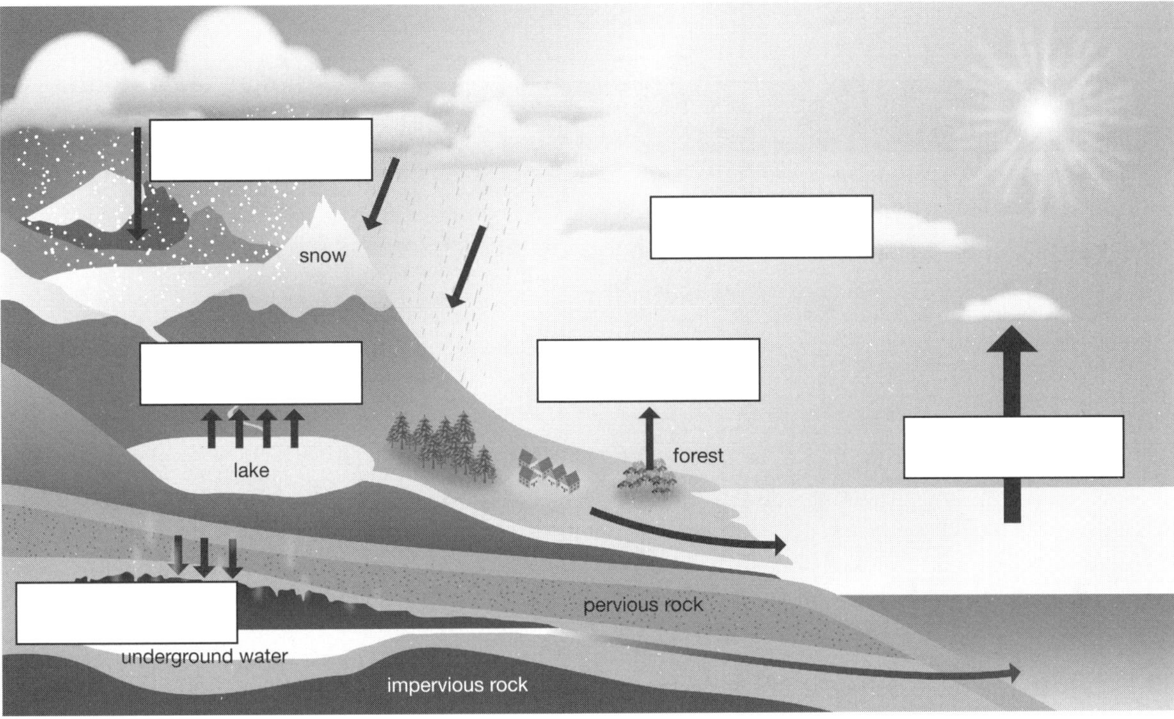

Figure 3.6.1 The water cycle

2 What is the source of energy that drives the water cycle?

___

3 **(a)** Identify a place in this cycle where the water would stay for a long time without changing state.

___

(b) Justify your response.

___

___

4 **(a)** Compare pervious and impervious rocks.

___

___

(b) Explain why underground water has collected where it is shown in the diagram.

___

___

3.6 The water cycle

(5) **(a)** In the table below, list changes to the water cycle that you think could occur in winter when conditions are very cold. Some examples are given.

(b) Justify your response in each case.

Change	Justification
• rate of movement of water slows down • rate of transpiration decreases	

(6) **(a)** In the table below, list changes to the water cycle that you think could result from very hot and humid conditions. Some examples are given.

(b) Justify your response in each case.

Change	Justification
• rate of evaporation decreases when humidity is high • high temperature increases the rate of evaporation	

RATE MY UNDERSTANDING
Shade the face that shows your rating   

3.7 Flowing water

Science inquiry skills

FOUNDATION	**STANDARD**	ADVANCED

Processing & Analysing

furrow (*n*) a long narrow trench made in the ground by a plough, used for planting seeds and plants

Farmer Jack lives in an area that gets a lot of heavy rain. He has been ploughing his land at the base of the hill and growing beans for years. He gets in his tractor and starts at one corner of the field and goes along the fence, turns and then goes down the field the other way.

He expanded his bean crop higher up the hill and used the same pattern of ploughing. Up the slope, turn, down the slope, turn and then up again. The pattern is shown in Figure 3.7.1. Is this the best way of ploughing the land?

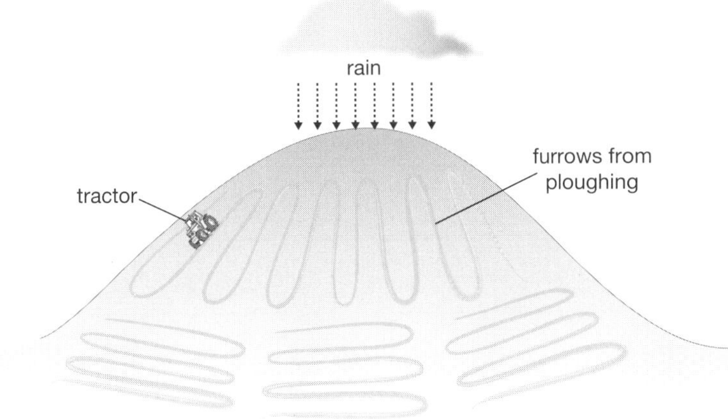

Figure 3.7.1

1. **(a)** Predict what will happen on farmer Jack's newly ploughed field when the next heavy rain falls.

   ____

   ____

   (b) Farmer Jack could probably manage his land better. Recommend changes he could make to achieve this.

   ____

   ____

In mountainous parts of the world, farmers have no flat land on which to grow crops. Over thousands of years terraces have been built, like those shown in Figure 3.7.2. The land has been cut away in places and built up in others to create narrow strips of flat land. On the outside edge of the land is a low wall.

These mountainous areas have a lot of rain at times but they also have strong winds.

Figure 3.7.2 A terraced mountain

2 (a) Predict the effect that strong winds would have on the moisture in the soil.

(b) What do you think would happen to the rain on this land if the mountain was not terraced?

(c) Describe what impact the terraces have when there is rain.

(d) Terraces have a low wall on the outside edge of each of the terraced steps. Explain the advantages of this wall.

(e) Which of the two mountainous landscapes—the terraced or the non-terraced—makes better use of natural resources? Give reasons to support your answer.

In Australia, sloping land that is used for grazing may have lines of mounded earth like those shown in Figure 3.7.3. These are not natural. They have been created by the farmers.

3 (a) What do you think is the advantage of creating these mounds on sloping land?

(b) Propose reasons why farmers created the mounds by cutting a trench and then filling it with pervious materials.

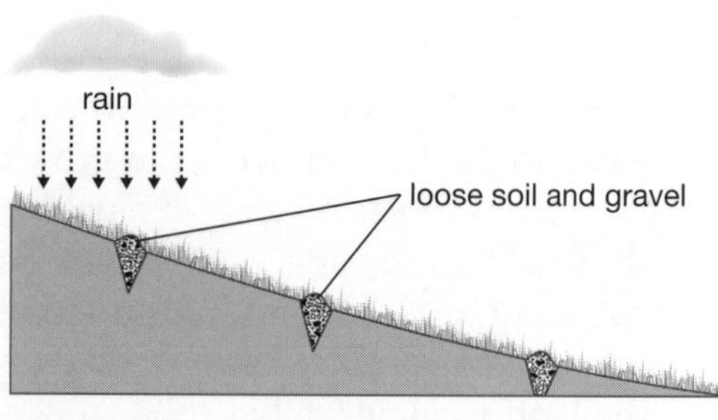

Figure 3.7.3 Grazing land with mounds

RATE MY UNDERSTANDING
Shade the face that shows your rating

3.8 Energy consumption

Science inquiry skills

FOUNDATION | **STANDARD** | ADVANCED

Processing & Analysing

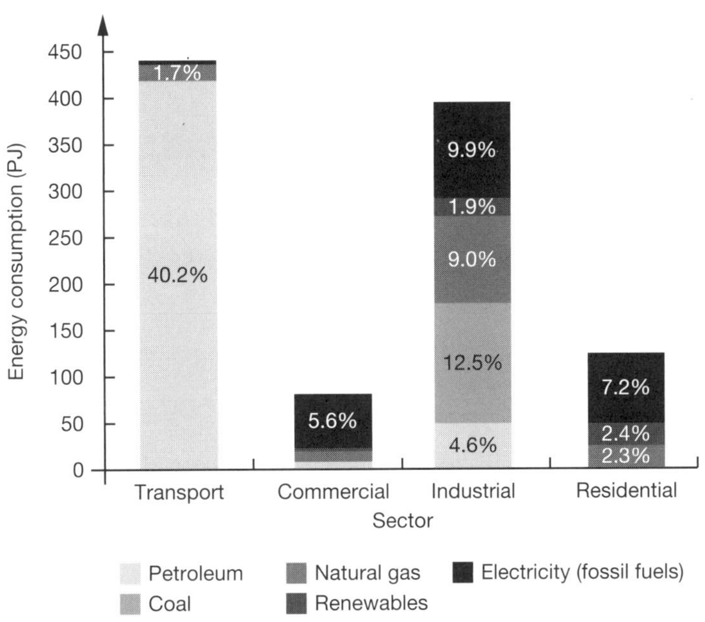

Figure 3.8.1 The breakdown of energy sources used to power New South Wales and the Australian Capital Territory energy needs as correct from 2007–8

The graph in Figure 3.8.1 shows the breakdown of energy use as a percentage of total energy consumption across the key sectors of transport, commercial, industrial and residential use in New South Wales and the Australian Capital Territory during 2007–8.

1 The energy quantities stated on the graph are given in petajoules (PJ).

1 PJ = 1 000 000 000 000 000 J. Explain why the vertical axis is stated in terms of such a large unit of energy.

____

2 List the four key sectors of energy use (transport, commercial, industrial and residential) in order from the sector that uses the greatest amount of energy to the sector that uses the least.

____

3 **(a)** State the energy source that is used to power most of the energy needs in the transport sector.

____

(b) Predict any changes that may happen to alter this breakdown of energy use in transport over the next 20 30 years.

____

____

3.8 Energy consumption

4 **(a)** Compare the breakdown of energy sources used in the commercial and industrial sector.

(b) List three examples of commercial businesses and three examples of industrial businesses.

(c) Explain how the activities of commercial and industrial businesses are different.

(d) Propose reasons to explain why the breakdown of energy use is different for the two sectors.

5 **(a)** Name the sector of energy use that corresponds to the energy you use in your home.

(b) State the major source of this energy.

(c) Explain why this is the major energy source at the present time.

(d) Propose ways in which this energy breakdown could change in the coming decades.

RATE MY UNDERSTANDING
Shade the face that shows your rating

3.9 Literacy review

Science understanding

FOUNDATION	STANDARD	ADVANCED

1 Use the following clues to find key terms from the chapter.

Clue	Word
process of breaking rocks down into smaller pieces	_ _ a t _ _ _ _ _ _
removal of sediments from one place to another	e _ _ _ _ _ _
special substances found in rocks	_ _ n _ _ _ l _
process by which plants use carbon dioxide, water and sunlight to make food	p _ o _ _ s _ n _ _ _ _ _ _
something that satisfies a particular purpose or need	r _ s _ _ _ _ _
wind, sunlight and tidal energy are all this type of energy source	_ _ _ _ w _ _ _ _
an energy source of limited supply is said to be	n _ n _ r _ _ _ _ _ _ _ _
oil, coal and gas are examples of these	_ o _ _ i _ f _ _ _ _
energy generated by falling water	h _ _ _ _ e _ _ _ _ _ _ _
biological material used as an energy source	_ _ _ m a _ _
device that converts sunlight into electrical energy	s _ _ a _ c _ _ _
energy source from beneath the earth's crust	_ e _ t _ _ _ _ _ _
evaporation of water from plants	_ _ a _ _ p _ _ _ _ i _ _
amount of water vapour in the air	_ u _ _ _ _ t _

2 Indicate whether the following statements are true or false by circling the correct answer.

Statements	Circle correct answer
Natural resources are non-renewable.	True / False
Cement, glass and medicines are human-made resources.	True / False
A renewable resource can be replaced by natural processes at about the same rate that it is being used.	True / False
Coal, oil and natural gas are renewable resources.	True / False
Land covers more of the surface of the Earth than water.	True / False
Animals breathe in oxygen and release carbon dioxide to the atmosphere.	True / False
Fracking is associated with mining.	True / False
Hydroelectricity and solar power are non-renewable energy sources.	True / False
Australia, Canada and the United States of America have the highest energy use per person of all countries.	True / False
Biogas is an alternative energy source made from the heat generated from the sun on hot rocks.	True / False
Geothermal energy relies on the heat from molten rocks under the earth.	True / False
Evaporation of water from plants is called transpiration.	True / False
Impervious rocks allow water to soak into them.	True / False
A rain garden is an example of water management that aims to filter stormwater and return it to natural waterways.	True / False

RATE MY UNDERSTANDING
Shade the face that shows your rating

3.10 Thinking about my learning

Write your own report for your parents about the work you have done in the chapter on Earth resources.

Student name: ______________________

Class: ______________

Subject: ________________

I learnt about ______________________

I learnt to ______________________

I demonstrated a very good understanding of ______________________

My level of effort and application was ______________________

My best piece of work was ______________________ because

I am good at ______________________

I need to work to improve ______________________

CHAPTER 4

Mixtures

4.1 Knowledge preview

Science understanding

FOUNDATION | **STANDARD** | ADVANCED

1 Read through the words in the vocabulary key below.

(a) List the words that you know and write a definition for each.

(b) At the end of this unit of work, review your original list of words and definitions. Add any new words you have learnt and write a definition for each.

aqueous solution	concentrated	dilute	dissolve
insoluble	mixture	saturated	solution
soluble	solute	solvent	suspension

2 Which of these substances is a mixture? (circle your answer)

A oxygen

B salt

C sea water

D pure water

(3) List some examples where we separate substances at home.

4.1 Knowledge preview

4 A mixture is best described as: (circle your answer)

A two or more substances chemically joined

B two or more substances that can be easily separated

C made up of only one substance

D a number of different substances mixed together but not chemically joined

5 Look at the pictures below and match the image, the separation technique and description to each picture.

Image	Separation technique	Separation description
Salt lake in Western Australia 	gravity separation	A sieve traps bigger particles but allows smaller particles to pass through.
Panning for gold 	filtration	Heavier substances sink to the bottom and lighter substances can be poured off.
Drip-filter coffee 	evaporation	Heat causes water to vaporise and large deposits of a solid substance remain.

6 Fresh water is essential to life on Earth. Explain where you think your tap water comes from.

____

____

____

____

HINT
Think about what happens between rain falling and water reaching your home.

4.2 Solubility and temperature

Science inquiry skills

FOUNDATION	STANDARD	ADVANCED

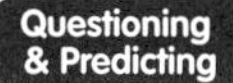

Maria and Tony did an experiment on solubility of substances at different temperatures. The aim was to find out if different substances have different solubilities when the temperature is changed. They tested how much copper sulfate would dissolve in water at different temperatures. They repeated the experiment with potassium sulfate. Table 4.2.1 shows their results.

Table 4.2.1 Results of solubilities of copper sulfate and potassium sulfate at different temperatures

Temperature (°C)	How much copper sulfate dissolves (g per 100 g water)	How much potassium sulfate dissolves (g per 100 g water)
0	22.3	7.4
10	27.2	9.3
20	32.3	11.1
30	37.8	13.0
40	44.8	14.8
50	52.8	16.5
60	62.5	18.2
70	73.4	19.8
80	87.5	21.4
90	105.4	22.9
100	124.9	24.1

(1) Plot both sets of data on the graph below. Label each line on the graph.

Results of solubilities of copper sulfate and potassium sulfate at different temperatures

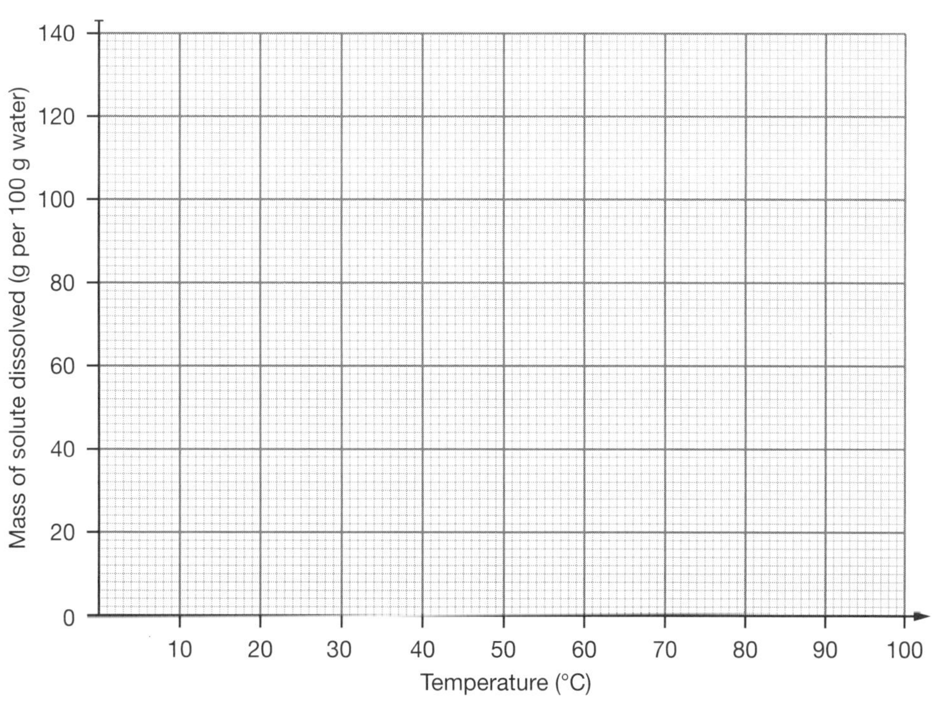

4.2 Solubility and temperature

2 In this experiment name the:

(a) solvent ______________________________

(b) solutes. ______________________________

(3) Indicate whether the following statements are true or false by circling your answer:

(a) The copper sulfate solution is more concentrated at 80°C than at 0°C. True / False

(b) The copper sulfate solution is more dilute at 10°C than 50°C. True / False

(c) The potassium sulfate is more dilute at 60°C than 10°C. True / False

(d) The less copper sulfate in the water, the more concentrated the solution. True / False

(4) State how much copper sulfate was dissolved in the water at the temperatures listed.

(a) 10°C ____________

(b) 80°C ____________

(c) 45°C ____________

(5) State how much potassium sulfate was dissolved in the water at temperatures listed.

(a) 30°C ____________

(b) 25°C ____________

(c) 100°C ____________

(6) Identify the effect that increasing the temperature had on the solubility of:

(a) copper sulfate

____

(b) potassium sulfate.

____

(7) Which of the two chemicals, copper sulfate or potassium sulfate, was more soluble at any temperature? Give evidence from the table or graph to explain your answer.

____

____

(8) Which of the two chemicals, copper sulfate or potassium sulfate, was more affected by increasing the temperature? Give evidence from the data to explain your answer.

____

____

RATE MY UNDERSTANDING
Shade the face that shows your rating   

4.3 Solubility of liquids

Science inquiry skills

FOUNDATION	STANDARD	ADVANCED

David and Susan did an experiment mixing different substances. The aim was to test how well different substances mixed together. They placed 5 cm of a particular solvent in a test-tube, then added 1 cm of a solute. They then shook them together to see if the solute dissolved. Their results are shown in Table 4.3.1 below. A tick means they mixed well. A cross means they did not seem to mix and separated out into two layers.

Table 4.3.1 Results showing how well different liquids mixed together

Solvent (5 cm in test-tube)	Solute (1 cm added into test-tube of solvent)					
	Kerosene	Water	Methylated spirits	Cooking oil	Glycerol	Turpentine
Kerosene	no need to do	✘	✘	✔	✘	✔
Water	✘	no need to do	✔	✘	✔	✘
Methylated spirits	✘	✔	no need to do	✘	✔	✘
Cooking oil	✔	✘	✘	no need to do	✘	✔
Glycerol	✘	✔	✔	✘	no need to do	✘
Turpentine	✔	✘	✘	✔	✘	no need to do

1 Match the terms in the left-hand column with the definitions in the right-hand column by drawing lines between them:

Term
cooking oil
glycerol
kerosene
solute
methylated spirits
solvent
turpentine

Definition
a substance that dissolves another chemical
a bad smelling solvent used as a fuel and cleaner of paint
a substance that dissolves into another substance
a paint thinner
fats from plants and animals such as butter, coconut oil, sunflower oil, margarine
a liquid used as a fuel in industry and in homes
a colourless, odourless sugar alcohol used as a sweetener

(2) Identify the solutes that dissolved in water.

____

(3) Identify the solutes that dissolved in oil.

____

(4) Identify the solvents that dissolved oil.

____

4.3 Solubility of liquids

5. Identify the solutes that did not dissolve in oil.

____

6. Identify two solvents for:

   (a) methylated spirits _______________

   (b) kerosene _______________

   (c) turpentine. _______________

7. Identify a solute that dissolves in:

   (a) glycerol and also in methylated spirits _______________

   (b) cooking oil and also in kerosene. _______________

8. State whether each of these mixtures would form a solution. Write 'yes' or 'no'.

   (a) kerosene in cooking oil _______________

   (b) water in glycerol _______________

   (c) turpentine in kerosene _______________

   (d) turpentine in water _______________

9. David and Susan concluded that oil and water seem to be different types of solvents, and that there seemed to be two different types of substances—those that dissolved in water, and those that dissolved in cooking oil.

   (a) List the substances that dissolved in water.

   ____

   (b) List the substances that dissolved in cooking oil.

   ____

   (c) State whether the substances that mixed well in water also mix well with each other. Explain your answer.

   ____

   ____

10. Propose a way in which Susan and David could make substances that did not mix well with each other combine better.

____

____

RATE MY UNDERSTANDING
Shade the face that shows your rating   

4.4 Mixtures

Science understanding

FOUNDATION	STANDARD	ADVANCED

homogenised (*adj*) milk that has been treated to prevent the solids and liquids from separating

sediment (*n*) solid particles that settle or rest at the bottom of a liquid

1 Classify the type of mixtures in the table below and identify the states of the substances in each.

Mixture	Type of mixture: solution, suspension?	Are its particles: solid in liquid, liquid in liquid, gas in liquid, gas in gas, or liquid in gas?
milk (FULL CREAM MILK Homogenised)		
diluted cordial (ready to drink) (LEMON CORDIAL Concentrated)		
air (PUMP 'EM UP)		
aerosol deodorant being sprayed (PONG OFF! Deodorant Aerosol)		
pure orange juice in bottle (100% PURE ORANGE JUICE; sediment)		
medicine in bottle (NO BURN ANTACID Shake Well)		

RATE MY UNDERSTANDING
Shade the face that shows your rating

4.5 How to filter

Science inquiry skills

FOUNDATION	**STANDARD**	ADVANCED

When filtering a substance you may want to keep the filtrate (liquid), the residue (solid), or both. This is known as *recovering* the substance. Table 4.5.1 below describes the procedure to follow.

Table 4.5.1 Procedures to recover substances

What do you want to recover?	What should you do to the beaker containing the mixture?
only the liquid (filtrate)	Let the solid settle before decanting the liquid into the filter.
only the solid (residue)	Stir the substance in the beaker to suspend as much of the solid in the liquid as possible, then pour the suspension into the filter. Use distilled water to wash any remaining solid out of the beaker into the filter paper.
both liquid and solid	Follow the same procedure as for recovering the solid.

1 Explain why you must keep the level of the liquid in the filter paper below the top of the filter paper in the funnel while you are pouring in the liquid to be filtered.

____

____

2 Explain why you must not touch the filter paper with anything (such as a pencil, your fingers or a glass rod) while the liquid is filtering through into the beaker.

____

____

3 What does 'recovering a substance' mean?

____

____

4 **(a)** Use the flow chart below to show the steps you would use to recover only the water from a sample of muddy water. Add additional steps if necessary.

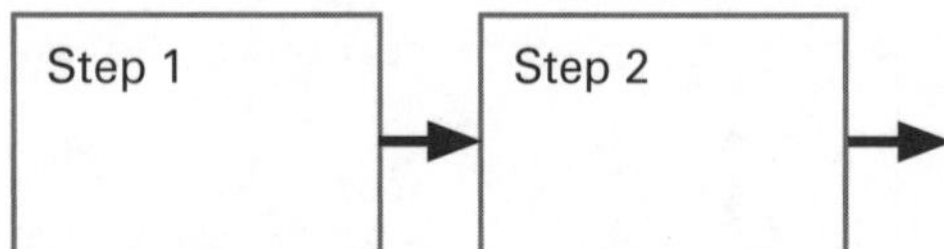

(b) Explain which way you would fold the filter paper to carry out this separation: conical or fluted? Explain why.

____

____

____

5 **(a)** Use the flow chart below to show the steps you would use to recover the solid and the liquid from a suspension of muddy water. Add additional steps if necessary.

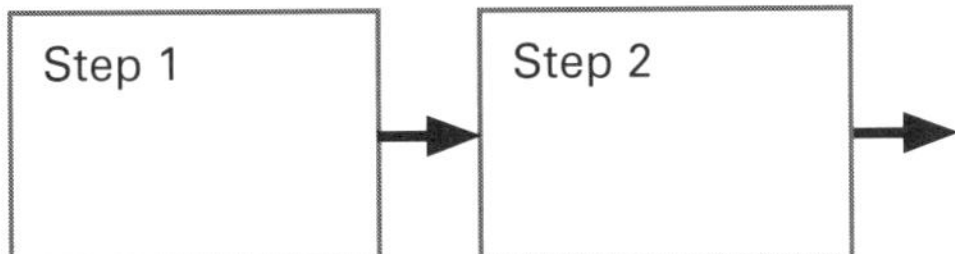

(b) Explain which way you would fold the filter paper to carry out this separation.

______

______

______

6 You want to recover all three substances in a mixture of water, silt (a soil finer than sand that becomes a suspension in water) and small stones (Figure 4.5.1).

Identify the method you would use to complete this task. Show each step in the flow chart below. Your method may be shown as text or as annotated scientific drawings.

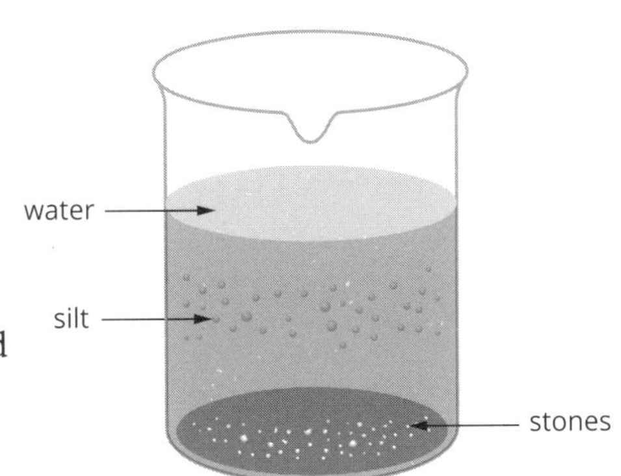

Figure 4.5.1

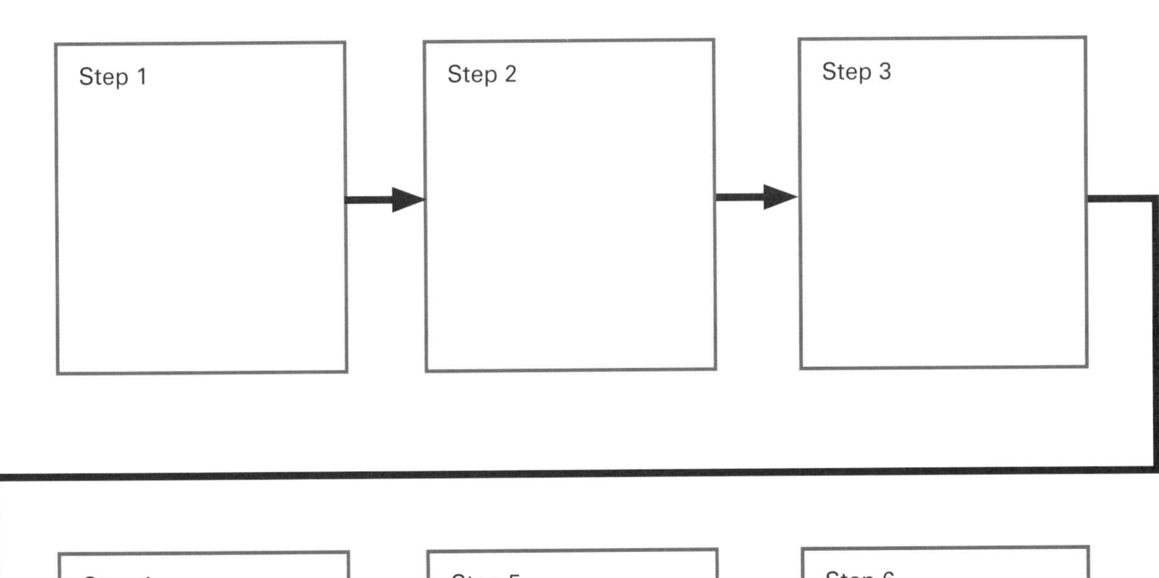

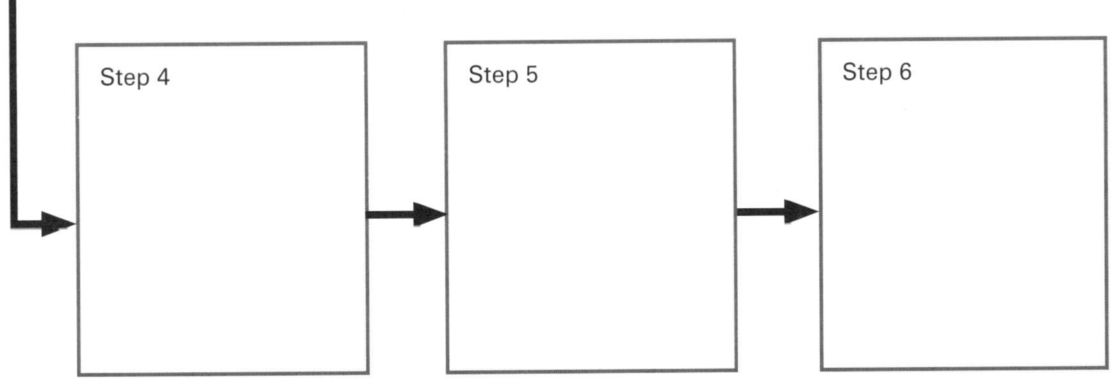

RATE MY UNDERSTANDING
Shade the face that shows your rating

4.6 Comparing filter papers

Science inquiry skills

FOUNDATION	STANDARD	ADVANCED

conical (*adj*) having a cone or funnel shape

fluted (*adj*) having ridges and narrowing at one end

Naomi and Glenn conducted an experiment to find out if the way a filter paper is folded affects how quickly it filters the mud out of a sample of very muddy water. They wanted to see how fast they could clean the water. Their teacher had shown them how to fold the filter paper into two different shapes—a conical fold and a fluted fold—which are shown in Figure 4.6.1.

The students set up a two beakers with a funnel in each one. They placed a conical filter paper in one funnel and a fluted filter paper in the other funnel. They poured 30 mL of the muddy water into each of two beakers and stirred them. Naomi then poured one of the 30 mL beakers of muddy water into the conical funnel, while Glenn poured the other 30 mL beaker of muddy water into the fluted funnel. They started their timers just as they started pouring the muddy water into the funnels. When all the water had stopped passing through the filter, they stopped their timers.

Naomi and Glenn knew that it is good practice to repeat experiments, so they did three trials for each type of filter paper. They recorded their results in a table as shown in Table 4.6.1. They then compared how long it took for all the water to pass though the two types of filter paper.

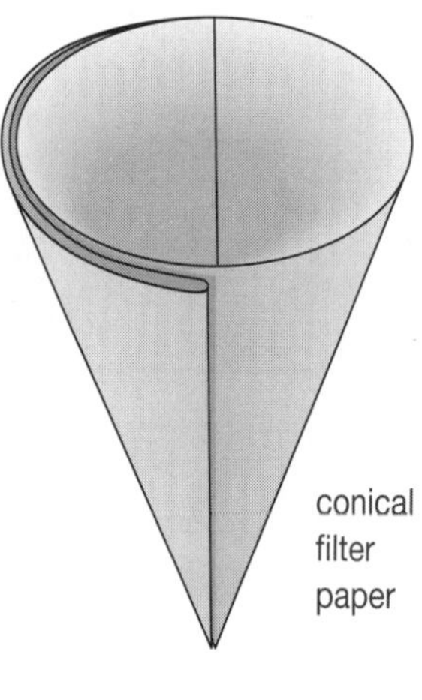

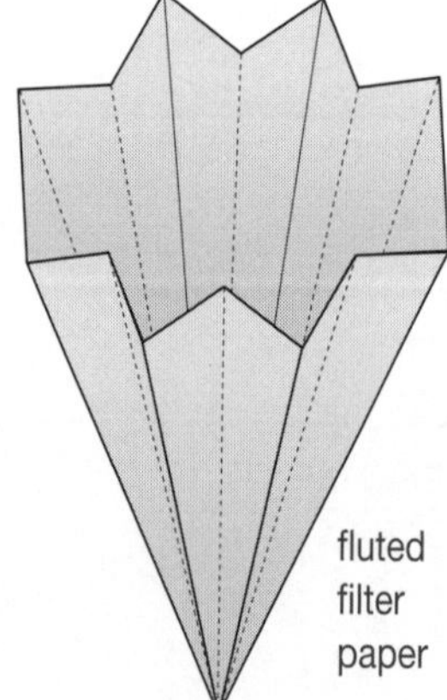

Figure 4.6.1

4.6 Comparing filter papers

Table 4.6.1 Time taken to filter 30 mL of muddy water using fluted and conical filter papers

Type of fold	Trial 1 (sec)	Trial 2 (sec)	Trial 3 (sec)	Average (sec)
Conical paper	55.4	62.2	56.6	
Fluted paper	37.6	42.1	38.4	

1. Show how to calculate the average time taken to filter with the conical shape filter paper in the space provided below. Enter your result in the table.

2. Show how to calculate the average time taken to filter with the fluted shape filter paper in the space provided below. Enter your result in the table.

3. For this to be a valid comparison of the two filter papers, list five variables that would have to be controlled.

4. Assess which paper cleaned the water fastest.

5. Explain the results.

RATE MY UNDERSTANDING
Shade the face that shows your rating

4.7 Separation of soluble substances

Science understanding

FOUNDATION	**STANDARD**	ADVANCED

There are various methods of separating soluble substances. They include distillation, evaporation, chromatography and crystallisation. The method used to separate soluble substances depends on what type of mixture you have. Distillation is used to separate a liquid from a solution. Evaporation is used to separate a soluble substance from a liquid. Chromatography is used to separate different-coloured dissolved substances. And crystallisation is used to separate a solute from a solution, then solidify the solution.

1. Students were instructed to separate a copper sulfate solution so they could recover both the water and copper sulfate. The separation method they used is shown in Figure 4.7.1.

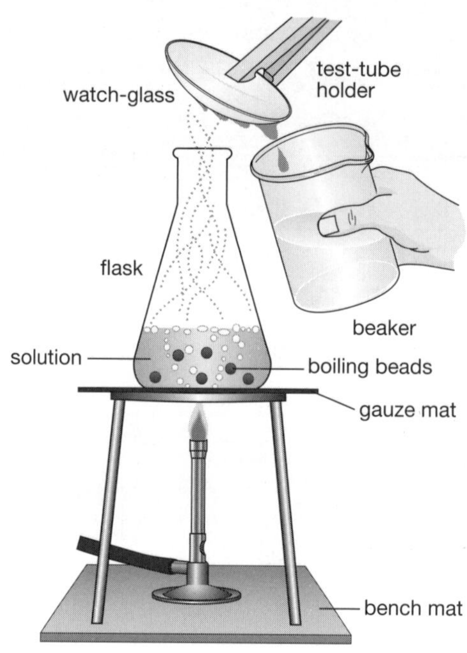

Figure 4.7.1

(a) Identify the main separation method used and explain how students collected liquid in the beaker.

____

(b) If the students continued to heat the conical flask, predict what it would contain at the end.

____

____

(c) Identify three possible safety issues with the experiment.

____

____

____

(d) How successful (efficient) do you think this method of separation would be? Discuss some problems with the process.

____

____

____

(e) Identify equipment that would do a better job of separating the copper sulfate solution to recover as much of the water and the copper sulfate as possible.

____

2 In another class, students were given a test-tube containing a mixture of sand and crushed mothballs (a white-coloured chemical called naphthalene which is used to kill insects). Their task was to separate the mixture of mothballs and sand, recovering both substances.

First, the students poured methylated spirits into the mixture in the test-tube. Then they warmed the test-tube by sitting it in a beaker of boiling water. The mothballs dissolved in the methylated spirits, but the sand did not. Once cooled, the students could see test-tube now contained sand and a colourless, clear liquid.

Next, the students filtered the contents of the test-tube through filter paper and collected the filtrate in a beaker. They made sure they rinsed out all the sand from the test-tube into the filter paper using some more methylated spirits. They then dried the filter paper in an oven. The last step was to put the beaker of filtrate on a hotplate and evaporate the liquid.

(a) Mothballs are insoluble in water. Are they soluble in methylated spirits?

____

(b) Propose why the students warmed the mixture in the text tube.

____

____

(c) Where would the naphthalene be in the mixture when the students are about to filter the mixture?

____

(d) What do you think they saw in the filter paper after they dried it?

____

(e) What probably appeared in the filtrate after heating it to evaporate the liquid?

____

(f) Name two methods of separation that were used to recover the solid mothballs and the sand.

____

RATE MY UNDERSTANDING
Shade the face that shows your rating  

4.8 Drinking recycled water

Science as a human endeavour

FOUNDATION	STANDARD	**ADVANCED**

Australia is one of the driest continents on earth. The low rainfall means recycled sewage water is sometimes used to irrigate crops and public parks. Shortages of drinking water can occur, especially during times of drought. The Australian Government has considered the use of recycled sewage for drinking water.

1 Your task is to work in pairs to research the use of recycled sewage for drinking water. Create six questions to guide your research using the questions starters what, where, when, why, who and how, as seen in Figure 4.8.1.

Some key ideas that could be included into your questions and research include:

- water treatment methods
- water quality
- issues or concerns about recycling sewage
- positives of recycling sewage
- global implications—how does it impact other countries?

Figure 4.8.1 Six types of questions to guide research

2 Find four different websites to source information on recycled sewage for drinking water. You should always check the usefulness and reliability and relevance of websites. Table 4.8.1 gives criteria to help you assess websites. Write the URL at the top of each column. Place ticks or crosses against each criterion as you evaluate the website.

Table 4.8.1 Website checklist

Reliable website criteria	URL 1	URL 2	URL 3	URL 4
I can identify who is responsible for the content on this site.				
The site has been updated in the last 3 to 6 months.				
This site was created by a credible person or organisation.				
The website has links to other credible websites.				
The links on the website lead you to other good information.				
The site has a .gov or .edu suffix.				
The site has useful contact information.				
The information on this site is similar to other sites I have found.				
The main purpose of this site is to provide facts (not opinions).				
Pictures on the site are helpful and have not been changed.				
Unreliable website criteria	**URL 1**	**URL 2**	**URL 3**	**URL 4**
The site is biased towards an opinion or point of view.				
The site contains spelling errors and broken links.				
The main purpose of the site is to sell a product.				
The site has no links to other credible websites.				

4.8 Drinking recycled water

What	Question:
	Information:
Where	Question:
	Information:
When	Question:
	Information:
Who	Question:
	Information:
Why	Question:
	Information:
How	Question:
	Information:

RATE MY UNDERSTANDING
Shade the face that shows your rating  

4.9 Types of mixtures

Science understanding

FOUNDATION | **STANDARD** | ADVANCED

Look at the diagrams and the key provided in Figure 4.9.1.

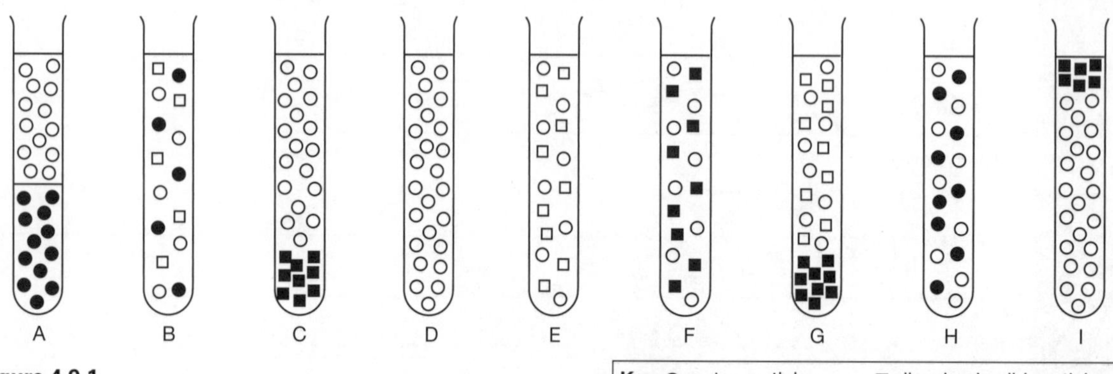

Figure 4.9.1

Identify the letter or letters that represent the test-tube containing:

1. a solution ________________
2. a suspension ________________
3. a saturated solution ________________
4. pure water ________________
5. a solution of one liquid in another ________________
6. an insoluble solid ________________
7. a mixture where both substances may be recovered by distillation ________________
8. a mixture that could be separated by filtration ________________
9. a mixture that could not be separated by filtration ________________
10. a mixture that could be separated by decanting ________________
11. the mixture most likely to be separated by centrifuging ________________
12. a solute best recovered by evaporation of water. ________________

RATE MY UNDERSTANDING
Shade the face that shows your rating   

4.10 Literacy review

Science understanding

FOUNDATION | **STANDARD** | ADVANCED

1 Recall key terms by drawing lines between each definition and its correct term.

Term	Definition
gravity	a lot of solute in the solvent
sterilisation	particles are spread throughout another substance but not dissolved
aquifer	separation method using difference in weight
concentrated	separation by pouring liquid off the top of a mixture of solid in liquid, or liquid in liquid
condenser	separation of solids or liquids from a liquid or gas by using a barrier with holes smaller than particles being excluded
dispersed	device that cools a gas and turns it into a liquid
filtration	a substance that will not dissolve in a particular solvent
insoluble	process of killing harmful microorganisms such as bacteria
decantation	layer of rock that contains water

(2) The crossword puzzle on mixtures has been completed but the clues are missing. Provide clues for each term on the crossword, by writing these in the space provided next to the puzzle.

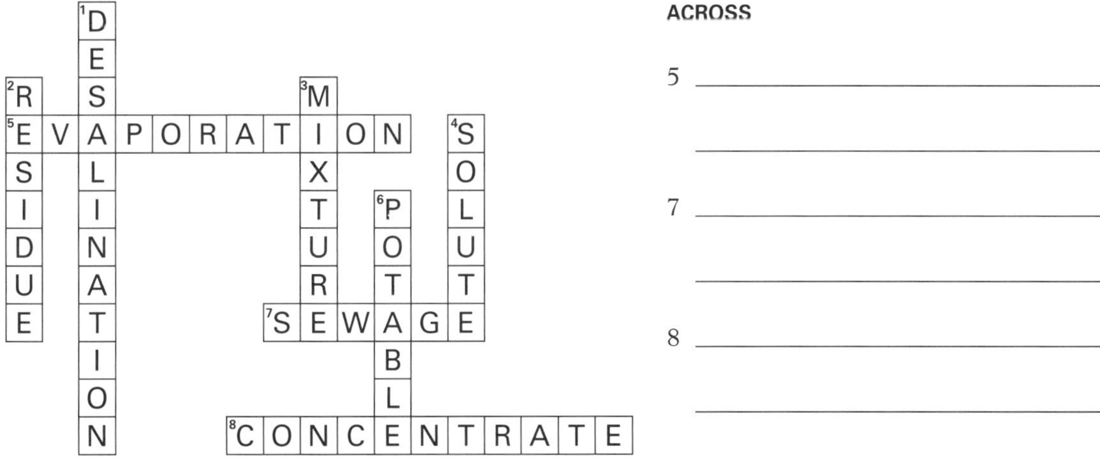

ACROSS

5 ______________________

7 ______________________

8 ______________________

DOWN

1 ______________________

2 ______________________

3 ______________________

4 ______________________

6 ______________________

RATE MY UNDERSTANDING
Shade the face that shows your rating

4.11 Thinking about my learning

Tick the square that best matches your understanding for each of the big ideas.

	Big ideas	I still need help with this	I understand this	I understand this well and can teach someone about this
Science understanding	I can recognise the differences between pure substances and mixtures.			
	I can give examples of pure substances and mixtures.			
	I can identify solvents and solutes in solutions.			
	I can recognise the difference between dilute, concentrated and saturated solutions.			
	I know what a suspension is.			
	I can explain the different techniques used to separate insoluble substances.			
	I can explain the different techniques used to separate soluble substances.			
	I can explain the different uses we have for water.			
	I can explain the different types of water treatment.			
Science inquiry skills	I can set up experiments involving separation techniques.			
	I can write experiment reports and discuss and explain my experiment results.			
	I can work safely in the science laboratory.			
Science as a human endeavour	I know and can discuss the issues relating to water use and management in the community.			
	I understand and can explain the decisions behind why we recycle greywater and blackwater.			
	I understand how sewage is recycled and can justify why we should/shouldn't use it for drinking water.			

CHAPTER 5

Habitats and interactions

5.1 Knowledge preview

Science understanding

FOUNDATION	STANDARD	ADVANCED

What do you know about habitats and interactions? Complete the multiple-choice quiz to check what you already know.

1 A habitat can best be described as the:

A way an organism adapts

B environment that sustains an organism

C interaction between organisms

D diversity of an ecosystem

2 The accurate list of abiotic environmental factors is:

A water, soil, bacteria, wind, moss

B water, bacteria, prey, rock, soil

C bacteria, prey, moss, predator

D water, soil, rock, wind

3 A food chain is:

A exactly the same as a food web

B a diagram that shows the feeding relationships and energy flows between organisms

C the interaction between two organisms where both benefit from the relationship and neither is harmed

D an animal that eats another animal

4 When a dingo eats a mouse, the mouse is:

A the prey and the dingo is the predator

B a first-order consumer and the dingo is a producer

C the predator and the dingo is the producer

D the prey and the dingo is third-order consumer

5 The organisms closest to extinction are called:

A vulnerable species

B rare species

C endangered species

D remnant species

6 In a sustainable ecosystem:

A living organisms would only have one food source

B living organisms could survive over short but not long periods of time

C all the needs of living organisms are not met

D living organisms are not threatened with extinction

7 These are introduced species and are not native to Australia:

A sheep, kangaroo, rabbit, prickly pear cactus, wattle tree

B sheep, rabbit, prickly pear cactus, cane toad, dingo

C kangaroo, dingo, cane toad, wattle tree, rabbit

D wattle tree, prickly pear cactus, kangaroo

8 Mutualism, commensalism and parasitism are:

A ways organisms interact with each other

B biological controls used to contain introduced species

C processes used by plants in photosynthesis

D causes of extinction of species

9 A herbivore is an animal that:

A eats only plants

B eats plants and animals

C eats only animals

D eats an animal that eats only herbs

5.2 Different living spaces

Science understanding

FOUNDATION	**STANDARD**	ADVANCED

1 In this forest there are many different places where animals can live. Identify some of the different places you would expect to find animals living.

____

____

____

____

(2) Compare the table of abiotic (non-living) factors at different places in the forest.

	At the top of the trees	Close to the ground
wind speed		
amount of light		
amount of gases such as oxygen		
effect of a heavy rain storm		

RATE MY UNDERSTANDING
Shade the face that shows your rating  

5.3 Daily changes

Science inquiry skills

FOUNDATION	STANDARD	ADVANCED

Processing & Analysing

A small pool of water on a coastal rock platform like the one shown in Figure 5.3.1 undergoes big changes in environmental conditions over 24 hours.

At low tide on a hot, sunny day, the water temperature in the pool can be quite high. Since hot water contains less oxygen than cold water, the amount of oxygen in the water decreases. The heat might also cause some of the water to evaporate, which would make the remaining water saltier. If it rained, the water in the rock pool would become less salty.

At high tide, the rock pool might be completely under water. The pool then becomes part of the ocean, and anything living in it would experience the strong ocean currents. The rock pool would experience the pounding action of the waves as the tide goes in and out. The temperatures recorded in a rock pool over one day are given in Table 5.3.1 below.

Figure 5.3.1 A rock pool on a rocky platform along the coast

Table 5.3.1 Temperatures recorded in a rock pool over one day

Time	Temperature (°C)	Time	Temperature (°C)
6 a.m.	16	1 p.m.	16
7 a.m.	17	2 p.m.	20
8 a.m.	23	3 p.m.	24
9 a.m.	29	4 p.m.	22
10 a.m.	34	5 p.m.	25
11 a.m.	35	6 p.m.	28
12 noon	15	7 p.m.	21

5.3 Daily changes

(1) Plot the data shown in Table 5.3.1 on the graph provided. Write a title for the graph.

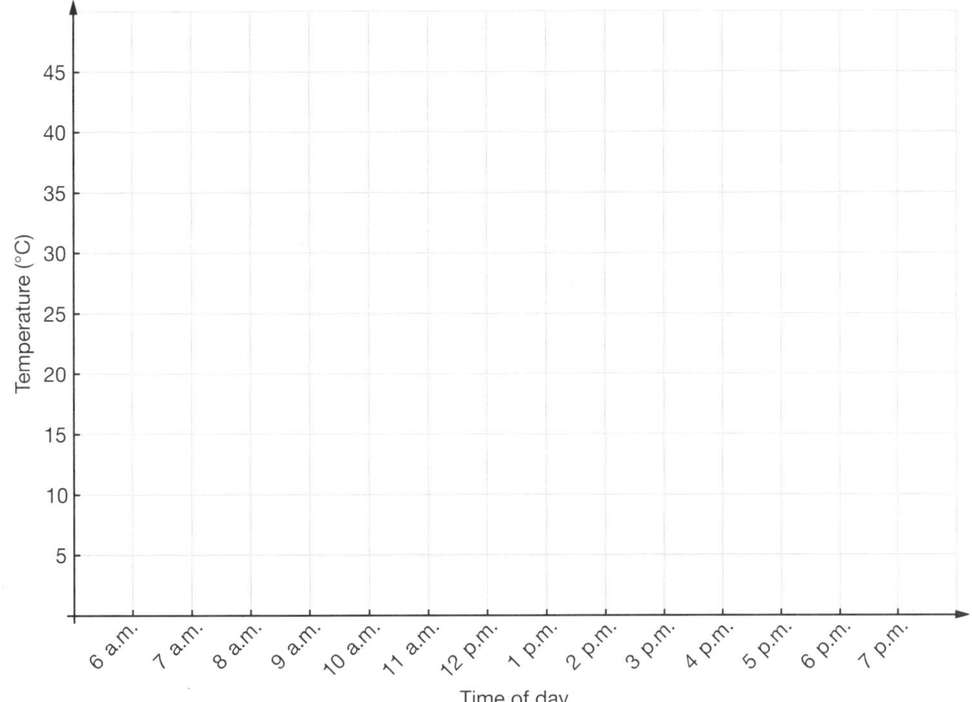

(2) Describe the environmental conditions that might explain the changes in temperature during the day.

(3) **(a)** Assume that you are an animal that is not able to leave the rock pool. What characteristics would you need to be able to survive in the rock pool?

(b) Justify the need for each suggested characteristic.

Characteristic	Justification

5.4 Adaptations

Science understanding

FOUNDATION | **STANDARD** | ADVANCED

All living things have special characteristics that help them to survive in their environment. These special characteristics are adaptations. Each of the pictures and descriptions below shows an organism that is adapted to its environment and highlights one particular adaptation.

organism (*n*) an individual animal or plant

1 Identify and describe the adaptation.

(2) Explain how the adaptation helps the organism to survive.

Lace monitors live in forests and woodland. They feed on insects, other lizards, small mammals, birds and birds' eggs.	
Seals need to be able to swim quickly to catch their prey and to escape from predators. Seals also come onto land.	
Eagles catch live prey with their talons and may carry it for long distances before they land somewhere to have their meal.	

RATE MY UNDERSTANDING
Shade the face that shows your rating

5.5 Living together

Science understanding

FOUNDATION	STANDARD	ADVANCED

1 Commensalism, parasitism and mutualism are three types of symbiosis—situations where organisms live together.

Identify each type of symbiosis by writing the name beside the description.

	Description	Type of symbiosis
(a)	An interaction between two organisms where both organisms benefit from the relationship and neither is harmed. The species may not be able to exist without each other.	
(b)	An interaction between two organisms where one of the organisms benefits and the other one is not affected.	
(c)	An interaction between two organisms where one organism gets food and shelter from the other organism and often harms or kills the other organism.	

(2) Identify each of the following as either commensalism, parasitism or mutualism.

(a) Head lice: Head lice are small, wingless insects that live and reproduce on the human scalp. They feed exclusively on human blood.

____

(b) Seed dispersal: Animals eat the fruits produced by plants to get the nutrients they need. The seeds of the fruit are not digested. They pass out with the animals' waste. The animals may carry the seeds a long way from the parent plants. When the seeds germinate and grow, the new plant is not competing with the parent plant for space, water and nutrients.

____

(c) Orchids and trees: Orchids are plants that grow on the trunks or branches of trees. They get the light they need as well as nutrients that run down the tree's branches. They do not affect the tree in any way.

____

(d) Coral reefs: Coral reefs are usually found in water where there are few nutrients for the coral polyps—the small animals that create the reef. Single-celled algae live inside the cells of the coral polyp. The algae are producer organisms able to make their own food. This food is shared with the coral polyp. The coral polyp provides a safe place for the algae to live.

____

RATE MY UNDERSTANDING
Shade the face that shows your rating   

5.6 Food chains

Science understanding

FOUNDATION	STANDARD	ADVANCED

swooping (*v*) moving quickly and suddenly flying downwards, possibly to attack

A food chain shows the feeding relationships in an ecosystem. In the following food chain, letters are used to represent the organisms.

A ⟶ B ⟶ C ⟶ D

1 Explain what the arrow in a food chain represents.

____

(2) Identify the letter in the food chain representing a:

(a) producer organism ________

(b) herbivore ________

(c) first order consumer ________

(d) carnivore. ________

3 Construct food chains from the following information.

(a) The hawk circled overhead before swooping down on the snake. Earlier that day, the snake had eaten a desert rat that was feeding on grass.

(b) A fat green caterpillar was munching on a leaf. It did not know that a butcher bird was perched on the branch above, ready to pick up the caterpillar in its hooked beak.

(c) Killer whales swim through the ocean in search of seals, which are their favourite food. In turn, seals search for large fish that have grown fat on their meals of smaller fish. Floating in the water are millions of plankton. These are microscopic producer organisms that provide food for small fish and other herbivorous animals in the ocean.

RATE MY UNDERSTANDING
Shade the face that shows your rating

5.7 Food webs

Science understanding

FOUNDATION | **STANDARD** | ADVANCED

herbivore (*n*) an animal that feeds on plants
organism (*n*) an individual animal or plant

Use the food web in Figure 5.7.1 to complete the following tasks.

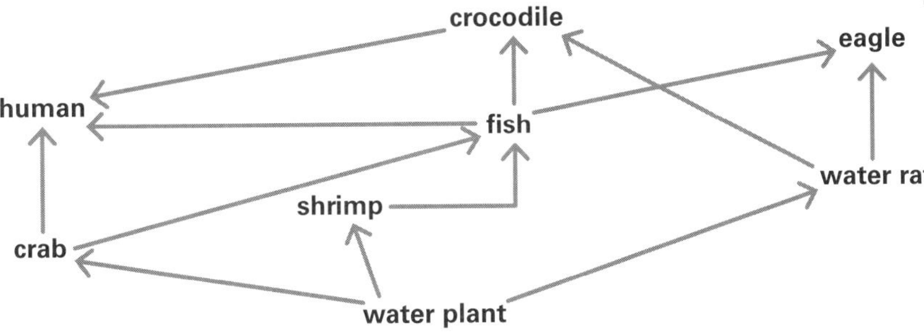

Figure 5.7.1

1. Construct two food chains from the food web.

2. Identify the herbivores in the food web.

3. Identify the organisms that are in competition for the water rat as a food source.

4. The human is a second order consumer when it eats the crab. Deduce what order of consumer the human is when it eats:
   (a) the fish ______
   (b) crocodile meat. ______

5. **(a)** What do you think would happen to the fish population if the water rat disappeared from the area?
   (b) Explain why you think this would happen.

6. **(a)** Based on your answers to question 5, predict what would then happen to the crab and shrimp populations if the water rat disappeared.
   (b) Give reasons for your answer.

RATE MY UNDERSTANDING
Shade the face that shows your rating

5.8 Creating a food web

Science inquiry skills

FOUNDATION | **STANDARD** | ADVANCED

Processing & Analysing

1. Food webs are formed by joining a number of food chains together. On the next page, construct a food web from the information in the following 10 food chains. They have many organisms in common. Each organism should appear only once in the food web.

grasses → insects → kookaburra

grasses → bandicoot → goanna

melaleuca tree → honey-eater → goanna

grasses → insects → frog → goanna

grasses → insects → frog → carpet snake

grasses → insects → insect-eating birds → goanna → carpet snake

grasses → insects → frog → water rat → goanna → carpet snake

grasses → insects → insect-eating birds → feral cat

grasses → bandicoot → feral cat

melaleuca tree → squirrel glider → feral cat

5.8 Creating a food web

Draw a food web in the box.

5.9 Human impacts in Antarctica

Science as a human endeavour

FOUNDATION	STANDARD	ADVANCED

Photographs of Antarctica show a land that appears untouched by human impact. However, in the 100 years that people have been travelling to Antarctica, they have left their mark. Past human activities that have harmed the ecosystems of the Antarctic and the Southern Ocean include fishing and hunting.

Hunting whales and seals

People began hunting for whales and seals in Antarctica in the early years of the nineteenth century. Within a few decades, the populations of these animals had decreased significantly. Some seal species, such as the Antarctic fur seal shown in Figure 5.9.1, were endangered. Eventually, seal hunting was stopped. Some of the islands off Antarctica became world heritage areas where the seals were protected. Since then, the seal populations have recovered and are no longer endangered.

Whaling was a very important industry in the Southern Ocean in the early 1900s. The number of whales reduced quickly. Falling profits drove many whaling companies out of business, and commercial whaling stopped in 1986. Whale populations appear to be recovering. Because whales live for a long time and have few young, it will take many years for their numbers to increase significantly.

Figure 5.9.1 The Antarctic fur seal is the smallest of the seals. It was hunted almost to extinction for its very thick, soft fur.

Fishing

Fishing still occurs on a large scale in Antarctica. As part of an international treaty to protect Antarctica, regulations aim to manage the industry and prevent over-fishing.

Reduction in the numbers of one species affects the predators and prey of that species and changes the dynamics of the food web. The regulations not only protect the fish that are wanted by fisheries—all other organisms in their food web are managed so that the environment is protected.

Research

The Australian Government, through the Australian Antarctic Division (AAD), aims to protect the Antarctic environment. The AAD employs scientists who investigate the effects of changes in the environment.

Krill are small shrimp-like creatures. The food web in Figure 5.9.2 shows that krill are food for a large number of species in the Antarctic food web. Scientists from the AAD are studying krill to learn how they are affected by biotic and abiotic factors in the environment.

Scientists from the AAD are also studying seal populations, many of the fish species and the seabirds that feed on the fish.

5.9 Human impacts in Antarctica

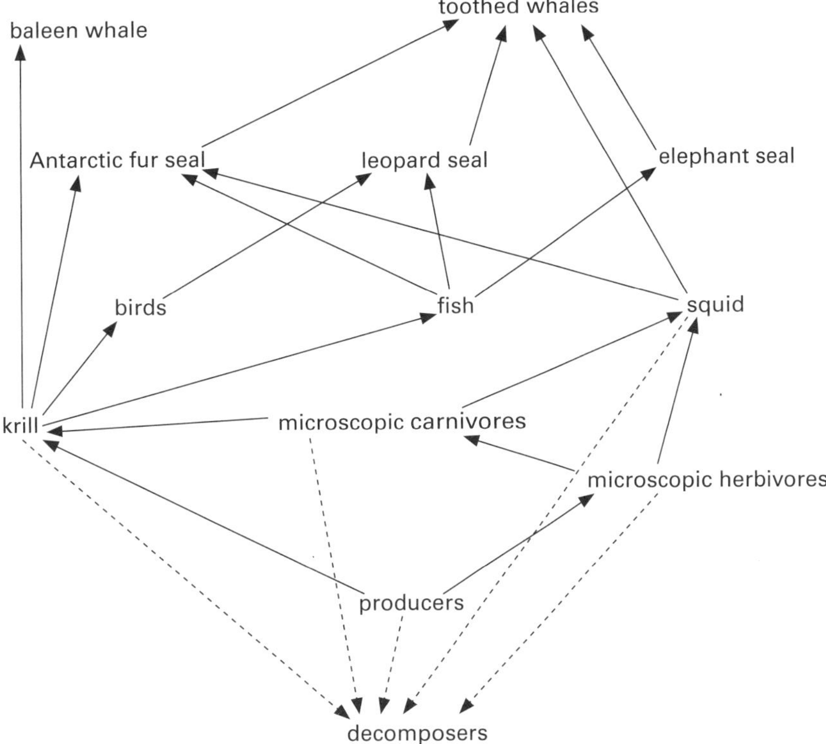

Figure 5.9.2 Antarctic food web

1 What has been done to protect the seal population?

____

____

2 Explain why it will take a long time for whale populations to increase.

____

____

3 Explain what Australian scientists are doing to protect the ecosystems of Antarctica.

____

____

(4) Use Figure 5.9.2 to identify the organisms that would be directly affected if the number of krill in the ocean decreased significantly.

____

____

5.9 Human impacts in Antarctica

(5) Identify any organisms in the food web that would not be affected at all if the number of krill decreased.

____

____

[6] Explain in your own words why krill are an important part of the Antarctic food web.

____

____

(7) Draw two food chains which include more than four organisms from the food web in Figure 5.9.2.

[8] Point out the positives and negatives of showing the interactions between organisms as a food web. Do this using the T-chart below.

Positives	Negatives

RATE MY UNDERSTANDING
Shade the face that shows your rating   

5.10 Growing crops

Science inquiry skills

FOUNDATION	STANDARD	ADVANCED

Processing & Analysing

When seeds are planted in the soil they will only germinate and grow if the soil temperature is suitable. The graphs in Figures 5.10.1 and 5.10.2 show the average soil temperatures in Hobart (Tasmania) and Perth (Western Australia). Table 5.10.1 contains information about the conditions required by certain crops for germination and growth.

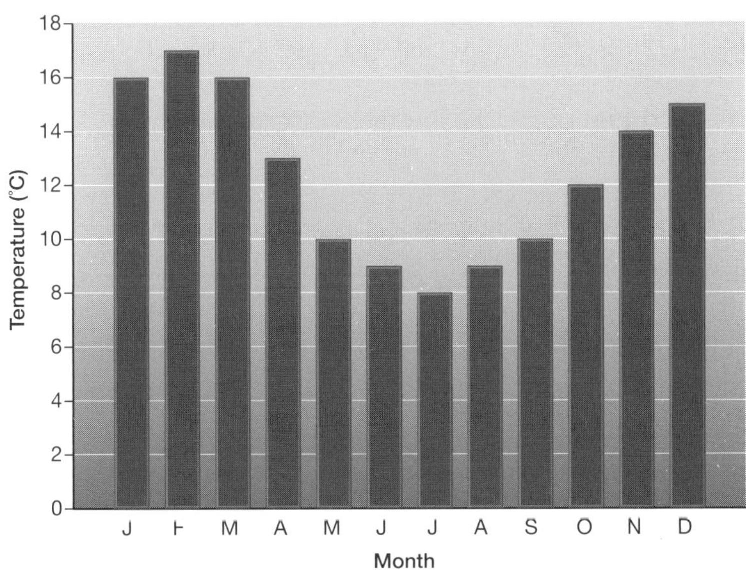

Figure 5.10.1

HINT

The scales used on the Y axis are different on the two graphs.

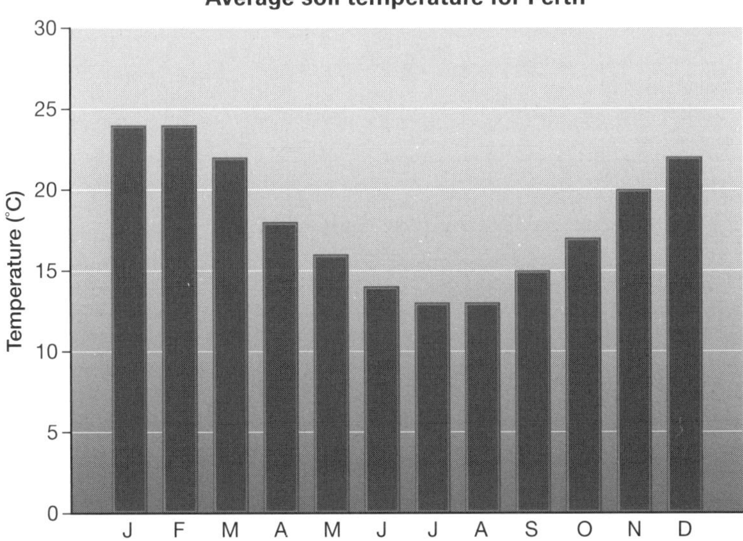

Figure 5.10.2

5.10 Growing crops

Table 5.10.1 Requirements of crops

Crop	Soil temperature range (°C)	Time for crop to mature (weeks)
brussels sprouts	7–30	12–13
capsicums	18–35	10–12
carrots	8–30	12–18
green beans	16–30	6–11
lettuces	8–27	8–12
rockmelons	20–32	10–16
watermelons	21–35	9–14
wheat	12–25	15–21

Use the information from Table 5.10.1 and Figures 5.10.1 and 5.10.2 to complete the following tasks.

1 **(a)** Identify the lowest average soil temperature in Hobart. ______________

(b) Identify the month in which it occurs. ______________

2 Compare the lowest average soil temperatures in Hobart and Perth.

____

3 If you were planting vegetables in August, name crops would you be able to plant:

(a) in Hobart ______________

(b) in Perth. ______________

4 **(a)** Recommend the best time to plant green beans in Hobart to ensure that you will have a mature crop. ______________

(b) Justify your answer.

____

5 Explain why capsicums can be grown more easily in Perth than in Hobart.

____

6 Explain why watermelons and rockmelons are not grown as commercial crops in Tasmania.

____

7 Describe ways that seasonal changes affect the type of farming that takes place in different parts of Australia.

____

RATE MY UNDERSTANDING
Shade the face that shows your rating   

5.11 Literacy review

Science understanding

FOUNDATION | STANDARD | ADVANCED

1 Write the terms that match the definitions below in the appropriate place in the word grid.

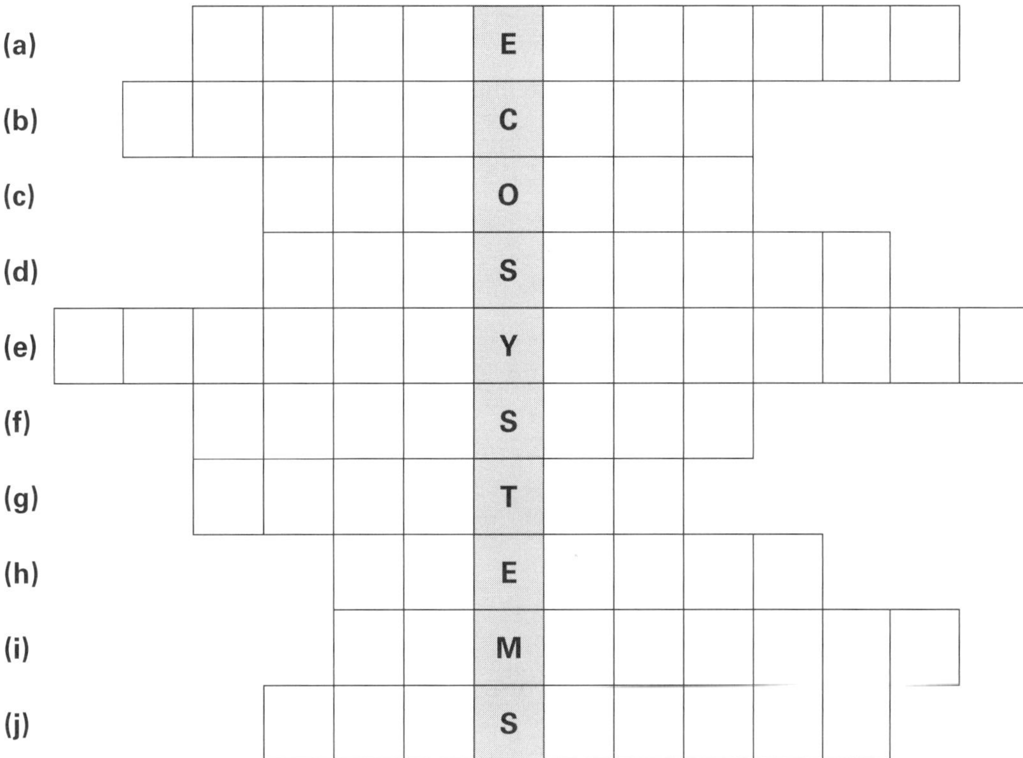

(a) Organisms that have the same food source and live in the same habitat.

(b) Organisms that are able to manufacture their own food.

(c) Non-living factors in the environment are described as this.

(d) Organisms that must eat other organisms to get the energy and nutrients they need.

(e) The process used by plants to make their own food.

(f) An organism that lives on or in a host, taking food or shelter from it.

(g) The place where an organism lives is called this.

(h) The term used to describe a group of living things with common characteristics.

(i) Another name for the interdependence of organisms.

(j) The place where all life as we know it exists.

RATE MY UNDERSTANDING
Shade the face that shows your rating

5.12 Thinking about my learning

Reflect on the chapter Habitats and interactions you have just completed. You may want to refer back to your science notebook to complete the activity.

Four things I have learnt...

1 ______

______

2 ______

______

3 ______

______

4 ______

______

Three interesting facts I learnt...

1 ______

______

2 ______

______

3 ______

______

Two things I can now do...

1 ______

______

2 ______

______

One question I still have/one thing I would like to know about:

______

Think–pair–share

Find a partner and discuss your response to 'One question I still have/one thing I would like to know about'. With your partner, try to answer the question or research what you would like to know more about. Fill in your information below.

______

______

______

______

CHAPTER 6

Classification

6.1 Knowledge preview

Science understanding

FOUNDATION	STANDARD	ADVANCED

1 What does classification mean?

____

____

2 Why do you think classification is important in science?

____

____

3 Below is a list of key terms that are commonly used in the study of classification in science. Look at these terms and complete the questions that follow.

amphibian	animal	aves	cells	class
conifers	dichotomous key	endoskeleton	exoskeleton	family
ferns	flowering plants	fungi	genus	invertebrate
mammals	marsupials	moss	phylum	reptile
seeds	species	taxonomy	vertebrate	weed

(a) Place a tick next to each term that you have heard before.

(b) Place a second tick next to each term that you can define.

(c) Select any five terms from the table. Use them in sentences to demonstrate that you understand their meanings.

____

____

____

____

____

____

6.2 Similar and different

Science inquiry skills

FOUNDATION	STANDARD	ADVANCED

Questioning & Predicting

1 Look carefully at these three critters. There are obvious similarities and differences.

Predict which two you think will be most alike: ______________________

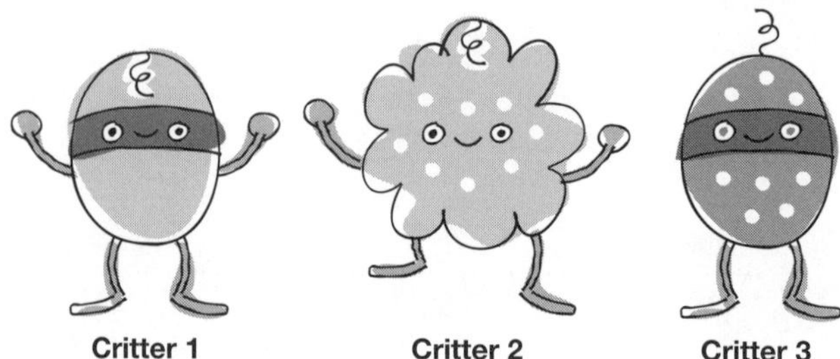

2 Observe the three critters. Compare them by identifying as many shared characteristics as you can. For each characteristic, list the critters that are similar. An example is provided.

Characteristic	Critter 1	Critter 2	Critter 3	Similar
arms	2 arms	2 arms	no arms	1 and 2

3 Calculate the number of points of similarity between the critters and enter your answers in the table below.

	Critters 1 and 2	Critters 1 and 3	Critters 2 and 3
Number of points of similarity			

4 State which critters are the most similar.

____

5 Is this what you predicted in question 1? State 'yes' or 'no'.

____

RATE MY UNDERSTANDING
Shade the face that shows your rating   

6.3 Identifying dogs

Science inquiry skills

FOUNDATION	STANDARD	ADVANCED

Processing & Analysing

dichotomous *(adj)* key provides two choices, with each choice leading to the next choice until the object is found

1 **(a)** The dogs shown in question 1(b) share some characteristics but not others. List characteristics that could be used to classify them.

(b) Use the key in Table 6.3.1 to identify the breeds of dog illustrated. Label each of the pictures when you have identified the dog.

Table 6.3.1 Dog breed dichotomous key

	Dog characteristics	Dog breed
1	**a** distinct black and white markings on face and body	Border Collie
	b not black and white	go to 2
2	**a** short-haired dog	go to 3
	b long-haired dog	go to 5
3	**a** large floppy ears	Dachshund
	b ears not floppy	go to 4
4	**a** short 'pushed-in' snout	Pug dog
	b long snout	Greyhound
5	**a** large floppy ears	Spaniel
	b ears not floppy	Maltese

RATE MY UNDERSTANDING
Shade the face that shows your rating  

6.4 Creating the key

Science inquiry skills

FOUNDATION	STANDARD	ADVANCED

Evaluating

Processing & Analysing

1. Here are six square people. Construct a dichotomous key to identify them.

2. **(a)** The following key was constructed to identify Carl, Harry, Mai, Steph, Karly and Jemma. This is a weak key. Explain why it is a weak key.

Characteristic dichotomous key			
1	**a**	boy	go to 2
	b	girl	go to 3
2	**a**	dimple in chin	Carl
	b	no dimple	Harry
3	**a**	light-coloured eyes	go to 4
	b	dark-coloured eyes	go to 5
4	**a**	striped dress	Mai
	b	plain skirt	Steph
5	**a**	hair in pony tail	Karly
	b	short hair	Jemma

(b) Suggest alternative characteristics that could be used to create a strong key.

RATE MY UNDERSTANDING
Shade the face that shows your rating

6.5 Scientific names

Science as a human endeavour

FOUNDATION | **STANDARD** | ADVANCED

When scientists discover a new plant or animal, they describe it and give it a name. Scientific naming of living things uses Latin and Greek. *Melaleuca viridiflora* is the scientific name for a plant with broad leaves. It is a paperbark tree native to the woodlands, swamps and streams of northern Australia. The first name (genus name) is a combination of two words: *mela* (black) and *leuca* (white). The bark of the Melaleuca is white with black areas. The second name (species name) is a combination of *virid* (green) and *flora* (flower). So *Melaleuca viridiflora* is a tree with green flowers that belongs to a larger group of trees that have white and black bark.

Naming animals

In Table 6.5.1 there is a small selection of words used in naming animal species. When words are combined, the last letter may be dropped. Sometimes letters such as 'e', 'i' or 'o' have to be added in the middle, or 'us' and 'um' added at the end.

Table 6.5.1 Words used in naming animal species

What it does		How it looks		Its body	
Word	**Meaning**	**Word**	**Meaning**	**Word**	**Meaning**
amphi	half	*aculeat*	pointed, spiny	*arctos*	bear
bates	walker	*cinereus*	grey	*canis*	dog
bios	life	*deinos*	terrible	*cheiro*	hand
carnis	meat	*erio*	woolly	*dactylo*	finger
gradus	step or walk	*fuliginosus*	sooty	*dent*	tooth
malus	bad	*lasios*	hairy	*hippo*	horse
odorus	smelling	*giganteus*	very large	*glossus*	tongue
perigrinus	wanderer	*macro*	large	*mastax*	mouth or jaw
potarno	river	*obesus*	fat	*placo*	flat
sulcata	digging	*nefrens*	toothless	*phascolo*	pouch
		rufus	red	*pteryx*	wing
		tachys	quick	*pus*	foot
		tri	three	*rhamphos*	curved beak
		verrucosa	covered in warts	*sauros*	lizard

1. The following four animals listed in the table below are all native to Australia: *Phascolarctos cinereus, Tachyglossus aculeatus, Macropus rufus, Macropus fuliginosus.* For each animal state what the name means and identify the animal you think it is.

Biological name	What it means	The animal is known as
Phascolarctos cinereus	pouched grey bear	koala
Tachyglossus aculeatus		
Macropus rufus		
Macropus fuliginosus		

2. If there was such an animal, propose what *Canis rufusobesus* would look like.

6.5 Scientific names

Naming plants

Table 6.5.2 lists some different words used to describe plants.

Table 6.5.2 Words and meanings used to describe plants

Word	Meaning	Word	Meaning
andr	male	*helio*	sun
brachy	short, little	*leuca*	white
calyptratus	caplike	*litho*	stone
callos	beauty	*mela*	black
carpa	fruit	*micro*	small
corne	made of horn	*stemon*	thread
eu	good	*teret*	rounded off, smooth
flora	flower	*truncatus*	cut off
folium	leaf	*viridis*	green

3 There are hundreds of different *Eucalyptus* trees in Australia. State what they all have in common that causes them to be grouped together.

Refer to Tables 6.5.1 and 6.5.2 for questions 4 and 5.

4 Descriptive words can be used for both plants and animals; for example, *Synanceia verrucosa* is the stone fish, *Hakea verrocosa* is an Australian native bush and *Lithops verrucoculosa* is a small plant growing close to the ground. Describe the characteristic they all have in common.

5 **(a)** If you were offered an *Odoromalasaurus* as a pet do you think you would take it?

Answer 'yes' or 'no' ______________

(b) Explain your answer.

6 Propose a good name for a:

(a) dog with grey, woolly hair ______________

(b) bottlebrush (*Callistemon*) tree with short rounded leaves.

RATE MY UNDERSTANDING
Shade the face that shows your rating

6.6 Arthropods

Science understanding

FOUNDATION | **STANDARD** | ADVANCED

Arthropods make up over 75% of all known animal species. All arthropods have a segmented body covered by an exoskeleton, and jointed limbs that enable them to move. Groups within the arthropods include the insects, arachnids, crustaceans, millipedes and centipedes.

Insects

Insects are the largest group of arthropods. Their body is divided into three parts—head, thorax and abdomen (Figure 6.6.1). They have a pair of antennae and a pair of compound eyes. Extending from the thorax are three pairs of legs for walking, jumping or digging. All insects include some individuals with wings at some stage in their life cycle. For example, young grasshoppers do not have wings, but they are present in the adult.

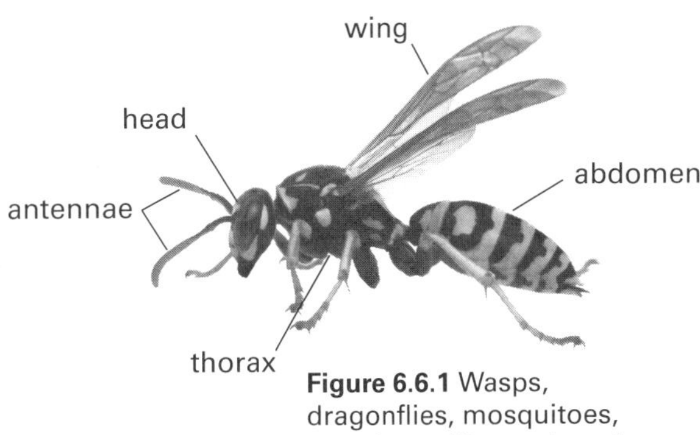

Figure 6.6.1 Wasps, dragonflies, mosquitoes, ants, butterflies and moths are insects

Arachnids

Arachnids have two body parts. The head and thorax are fused to form a cephalothorax (see Figure 6.6.2). Arachnids have four pairs of walking legs but do not have antennae. Spiders have fangs they use to capture their prey. Poison injected from the fangs paralyses or kills the prey. Many arachnids make webs to catch prey. The silk is produced by spinnerets at their tail.

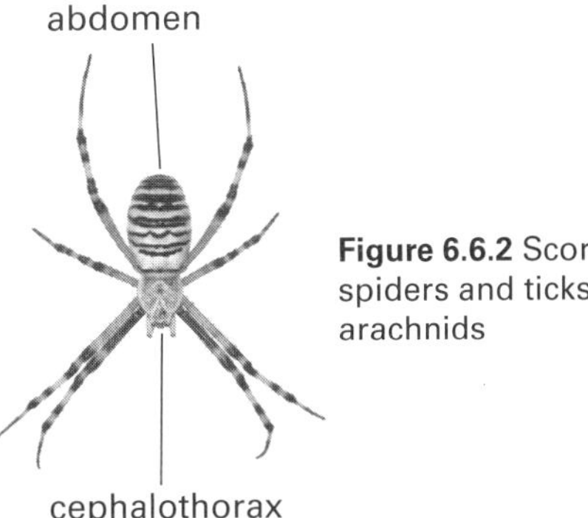

Figure 6.6.2 Scorpions, spiders and ticks are all arachnids

Crustaceans

Most crustaceans live in water. They have a cephalothorax. Crustaceans have two pairs of antennae and usually five pairs of legs (see Figure 6.6.3).

Sometimes the front legs are modified as pincers, which are used to catch and hold prey, and as protection.

Lobsters and wood lice are crustaceans, as are prawns and crabs.

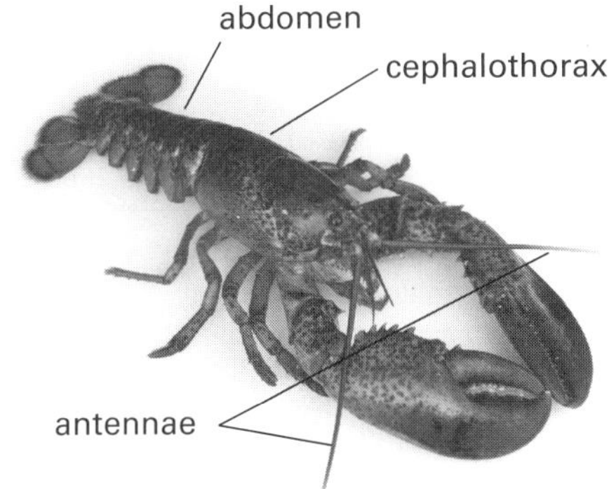

Figure 6.6.3 Lobsters, wood lice, prawns and crabs are crustaceans

Centipedes and millipedes

Centipedes and millipedes have legs on most of their segments. Centipedes have one pair of legs on each segment and are very fast moving. Millipedes have two pairs of legs on each segment. Although they have more legs, they are slow-moving animals.

1 Identify the centipede and millipede by writing the correct name in the boxes.

6.6 Arthropods

Exoskeleton

Having an exoskeleton has advantages and disadvantages. The skeleton provides protection and the waxes it contains make it waterproof, preventing the organism from drying out. This feature enables arthropods to live on land.

To allow movement, the exoskeleton has joints in the legs, antennae and between the segments of the body. At the joints the exoskeleton is thinner and more flexible.

Just like you, arthropods move by using muscles that act in pairs. Your muscles are attached to the outside of your skeleton. Arthropod muscles are attached on the firm ridges or bars inside the exoskeleton as shown in Figure 6.6.4.

Arthropods shed their skeleton or moult when they need to grow. You can see the damsel fly shedding its skeleton in Figure 6.6.5. Material to make a new exoskeleton is produced before the old exoskeleton splits, releasing the arthropod. While the new exoskeleton is still soft, the arthropod puffs itself up, making its body as big as possible. This stretches the new exoskeleton before it hardens. When it is moulting, the arthropod is very vulnerable to attack by predators so it stays hidden.

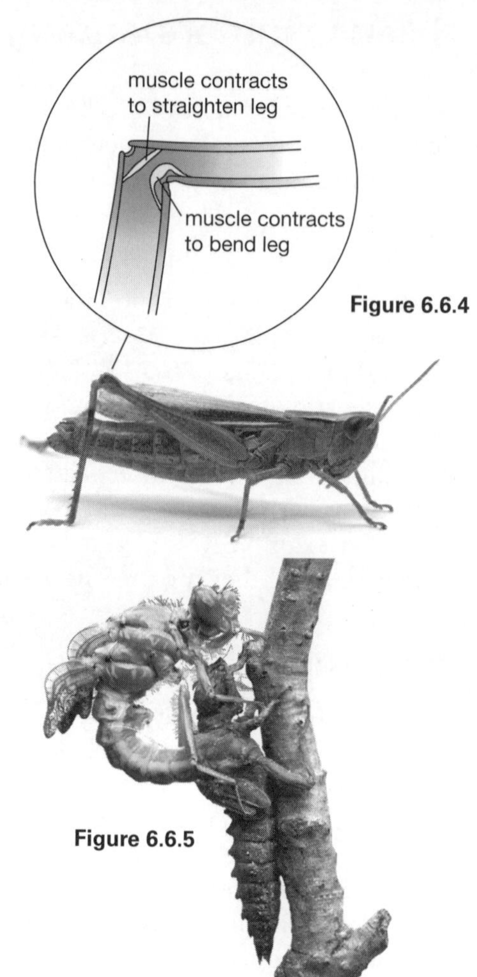

Figure 6.6.4

Figure 6.6.5

2 Identify and label the body parts of the butterfly in Figure 6.6.6.

3 Identify and label the body parts of the scorpion in Figure 6.6.7.

4 Identify and label the body parts of the woodlouse in Figure 6.6.8.

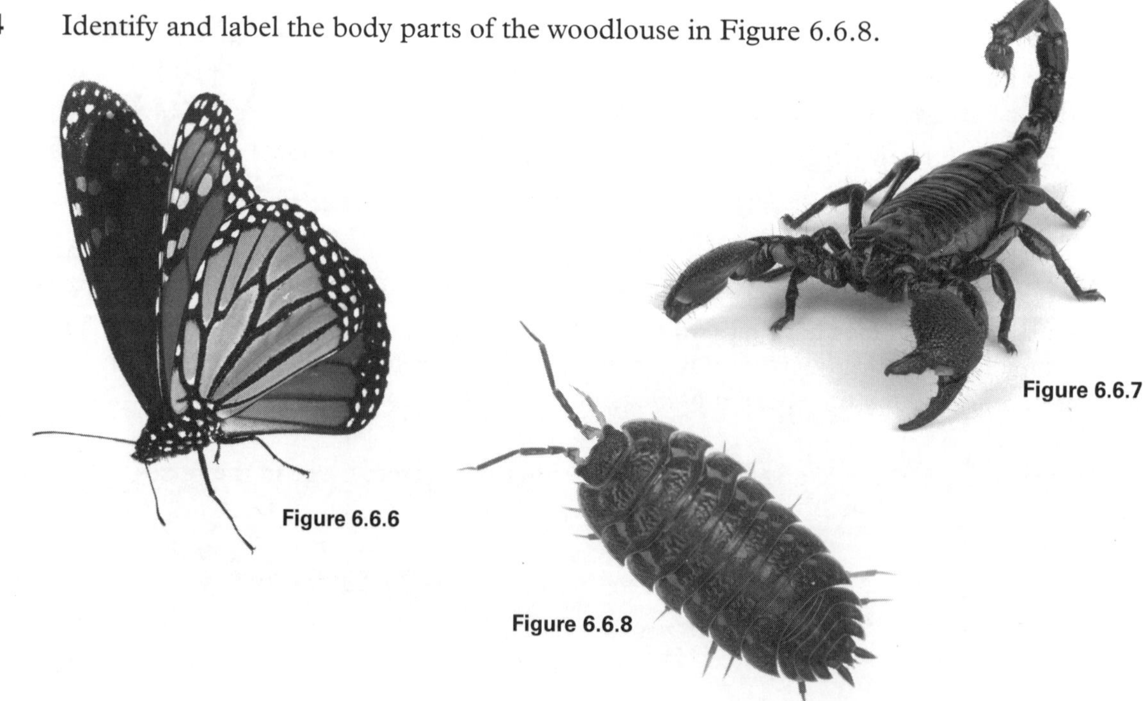

Figure 6.6.6

Figure 6.6.7

Figure 6.6.8

6.6 Arthropods

5 What characteristics are common to all arthropods?

(6) Complete the Venn diagrams below to:

(a) compare insects and arachnids **(b)** compare centipedes and millipedes.

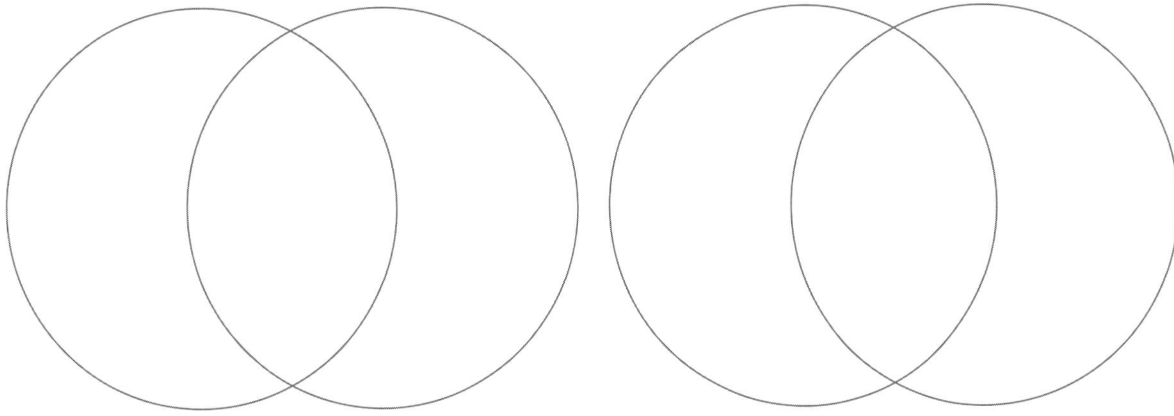

(7) Explain why it is important for arthropods to be able to moult.

(8) Explain why moulting is dangerous for arthropods.

9 Explain how arthropods make sure that there is room for growth in their new skeleton.

(10) Compare the way your leg bends with the way arthropods bend their legs.

RATE MY UNDERSTANDING
Shade the face that shows your rating

6.7 Doctor Joyce Vickery

Science as a human endeavour

FOUNDATION	STANDARD	**ADVANCED**

Doctor Joyce Vickery (1908–79) was one of Australia's leading botanists (Figure 6.7.1). She was born and educated in Sydney. When she graduated from The University of Sydney in 1935, she went to work at the National Herbarium of New South Wales, Royal Botanic Gardens. She spent most of her working life there. She was the first woman to be appointed as a scientific professional officer in the NSW Public Service.

Dr Vickery strongly supported conservation of natural environments and established campaigns to set aside state parklands. She was instrumental in establishing parklands that are now known as the Kosciusko National Park and the Berowra Valley Regional Park in the northern suburbs of Sydney.

Figure 6.7.1 Dr Joyce Vickery

POLICE DEPARTMENT,
SYDNEY.
8th July, 1960.

KIDNAPPING

Have You Seen This Boy?

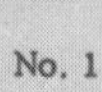

No. 1.

No. 2.

Your co-operation is urgently sought in locating GRAEME FREDERICK HILTON THORNE, 8½ years of age, who was kidnapped in the vicinty of his home in Edward Street, Bondi at about 8.30 a.m. on Thursday, 7th July, 1960, whilst on his way to school.

The boy is described as 8½ years of age, tall for his age, fair hair well oiled and making it appear brown, fair complexion, heavy build, dressed in grey Scots College uniform with facings of the College colours, being navy and gold, short trousers, light blue shirt, College tie, grey pullover, long socks, black shoes, grey serge cap with Scots College badge of unicorn on front. The uniform worn by Thorne is the same as that shown in photograph No. 2 above.

If you have seen this boy since 8.30 a.m. on Thursday, or have any information of his possible whereabouts, please contact the Police by ringing telephone 2222 or B030 or contact your local Police direct. A GOVERNMENT REWARD OF £5,000 HAS BEEN OFFERED IN CONNECTION WITH THIS MATTER.

C. J. DELANEY,
Commissioner of Police.

V. C. N. Blight, Government Printer

Figure 6.7.2 Kidnapping notice for missing boy Graeme Thorne, 1960

On many occasions, Dr Vickery used her botany skills to help the police solve crimes. Her best-known case was in 1960 before there was DNA analysis and other sophisticated forensic techniques.

Graeme Thorne was the 8-year-old son of the winner of Australia's first big lottery. A month after the win, Graeme was kidnapped and a ransom was demanded. Before it could be paid, Graeme's body was found in bushland. The police had a suspect.

They took tiny fragments of plants found in the rug the body was wrapped in and from the back of the suspect's car. Dr Vickery was able to demonstrate that the plant fragments on the rug came from bushes growing at the suspect's house. She also demonstrated that his car contained traces of plants found growing near where the body had been dumped.

6.7 Doctor Joyce Vickery

The suspect was charged and found guilty of the murder largely on the basis of Dr Vickery's analysis of plant material and soil at the crime scene.

Joyce Vickery was a taxonomist. Taxonomy is a branch of science which looks at the classification of organisms. When she started working in the area, this aspect of science was not highly valued. Through her dedication she turned it into a well-respected science.

She was often asked how you can tell when one plant is different enough from another to be recognised as a separate species. Dr Vickery responded that it is important to observe the detail of the structure of the leaves and stems, to consider the distribution across the world, to look at inherited characteristics and to learn about the history of the plant. However, Dr Vickery also suggested that many taxonomists develop an intuitive feel for their work based on a deep understanding of all aspects of the plants they are studying.

1 What field of work was Dr Vickery engaged in and where did she spend most of her working life?

____

____

2 What facilities do Australians have today, thanks to the work of Dr Vickery?

____

____

3 Explain how Dr Vickery was able to use her knowledge of plants to help the police in their investigations.

____

____

____

(4) What characteristics does a person need to be a good taxonomist?

____

____

[5] **(a)** Dr Vickery also suggests that a good scientist needs to have a characteristic that is rarely mentioned. What was this characteristic?

____

____

(b) Why do you think this characteristic would be helpful to a scientist?

____

____

RATE MY UNDERSTANDING
Shade the face that shows your rating   

6.8 Classification inquiry task

Science inquiry skills

FOUNDATION	**STANDARD**	ADVANCED

Processing & Analysing **Evaluating**

Key vocabulary:
class
dichotomous key
family
genus
invertebrates
kingdoms
order
phylum
species
taxonomy
vertebrates

The Earth has millions of different species of plants and animals. Scientists group organisms based on characteristics they have that are similar to other organisms. Scientists place these organisms into groups and give them species names. This makes it easier to communicate about and identify them.

Your task is to work in a group of 3 or 4 and brainstorm a list of native Australian animals. Write your animals on the brainstorming template below.

Brainstormer

1 Choose two of the animals from your brainstorming session and write their names at the top of the columns in Table 6.8.1. Next, list some characteristics of each organism in the left-hand column and place a tick or a cross under the name of each animal to indicate if it has this characteristic.

Table 6.8.1 Animal characteristics research findings

Characteristic	Animal 1: ______________	Animal 2: ______________
e.g. lives on land		
e.g. has four legs		

(2) Comment on the similarities and differences of the two organisms you have chosen.

(3) Choose four of the animals from your brainstorm list and create a dichotomous key, like the one below. Write the name of each animal in the boxes and the characteristic used to divide them into different categories outside of the boxes, next to the arrows. The aim is to have one animal in each box of the bottom row.

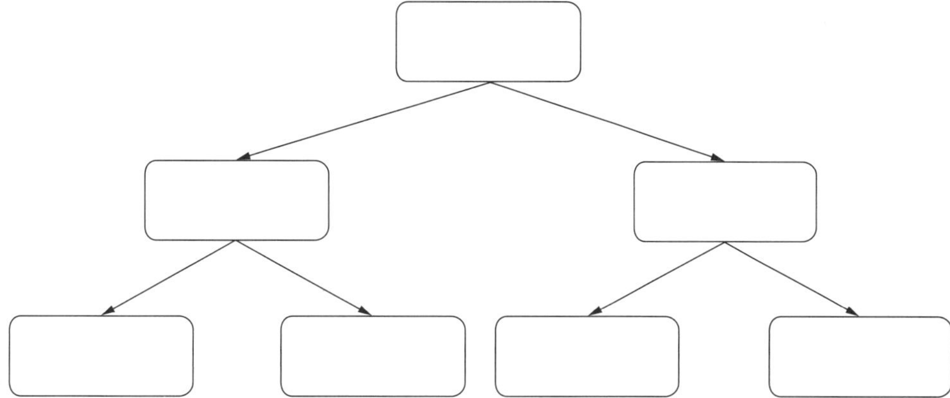

(4) **(a)** Repeat the dichotomous key for four different animals.

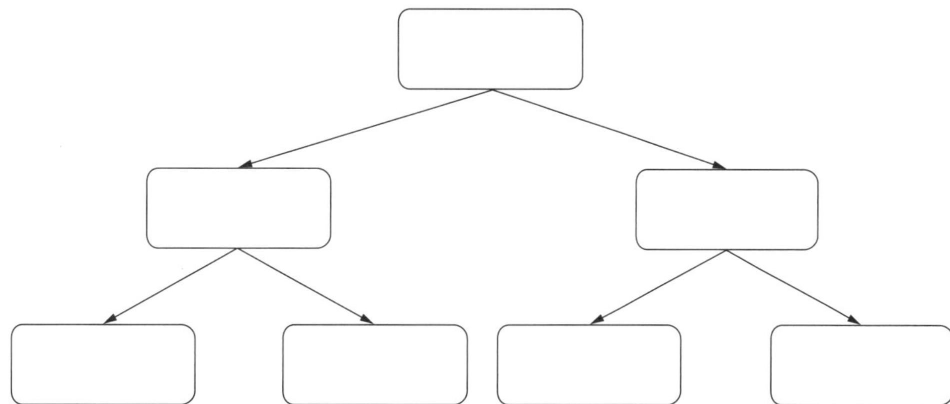

(b) Which is the better key? Explain.

(5) If a new plant or animal species were discovered, how do you think scientists could modify a classification key to include the new species?

RATE MY UNDERSTANDING
Shade the face that shows your rating

6.9 Literacy review

Science understanding

FOUNDATION	STANDARD	ADVANCED

1 A dichotomous key is used by scientists to identify items. The word dichotomous originates from Ancient Greece. Look at the break-down of the word dichotomous on the right.

dicho → **in two**

tomous → **to cut**

(a) In your own words, write a definition of the word dichotomous.

____

____

(b) Highlight the words below that convey the same idea as the term dichotomous.

add together	combine	divide	mix together
pull apart	separate	split	unite

2 Below is an example of a dichotomous key for identifying messages presented as a table.

	Message characteristic	Type of message
1a	is written	go to 2
1b	is verbal	go to 3
2a	is fast	email
2b	is slow	letter
3a	face-to-face	speaking
3b	over long distances	telephone

(a) Explain what features of this table make it a dichotomous key.

____

____

(b) In the space provided below, present the dichotomous table key above, as a flow chart key.

3. Recall the key terms of classification by matching them with their definitions. Draw a line from the key term to its definition.

Term	Definition
species	A skeleton inside the body
endoskeleton	Animals with a nerve cord running down their backs and an endoskeleton
taxonomy	A class that includes all the animals that have a body covering of hair and feed their babies on milk produced by the mother
mammal	The last level of classification of living things
dichotomous key	The science of grouping and naming things
exoskeleton	One of five kingdoms of living things. Multicellular or unicellular organisms with eukaryotic cells. A protein-rich cell wall is the outer layer.
fungi	A skeleton on the outside of the body
chordates	One of five kingdoms of living things. Multicellular organisms. Cells with a distinct nucleus. A cellulose cell wall is the outer layer.
plants	One of five kingdoms of living things. Multicellular organisms. Cells with a distinct nucleus. Has a membrane as the outer layer.
protists	One of five kingdoms of living things. Single-celled organisms with a distinct nucleus.
animals	Key with two choices at each stage
monerans	Animal with an exoskeleton and jointed limbs
arthropod	Describes animal with a body temperature that varies with the temperature of their surroundings
ectothermic	One of five kingdoms of living things. Single-celled organisms with a distinct nucleus.

RATE MY UNDERSTANDING
Shade the face that shows your rating

6.10 Thinking about my learning

What have you learnt about classification? Reflect back on the Activity Book worksheets you have completed and fill in the table below.

Reflecting on my learning	Reflecting on my thinking
I now know ...	I think ...
I learnt ...	I wonder ...
I liked ...	My question is ...
I can ...	I remember ...
It was hard to ...	I understand ...

CHAPTER 7

Forces

7.1 Knowledge preview

Science understanding

FOUNDATION | **STANDARD** | ADVANCED

1. Look carefully at the list of science terms in the box below. Most of the terms relate to the topic 'forces' but some are not associated with forces.

animal kingdom	balance	battery	chemical reaction
collision	force field	food web	friction
gravity	inertia	lever	magnet
mass	motion	newton	planets
positive charge	pulley	push	soil erosion
speed	spring	weight	

(a) Which of the areas of science does the study of forces belong to: chemistry, biology, physics, earth science, psychology or astronomy?

____

(b) Use a pencil to cross out all the terms that are not directly relevant to the study of forces.

(c) Look at each of the words in the box that relate to forces. Demonstrate the meaning of each of these words by using them in a sentence. You may include more than one of the words in the same sentence.

____

____

____

____

____

____

____

____

____

____

7.2 What can forces do?

Science understanding

FOUNDATION	**STANDARD**	ADVANCED

Forces may be pushes, pulls or twists. A force can be represented by an arrow pointing in the direction it is acting—the bigger the arrow, the bigger the force. Many forces can act on an object at the same time. If these forces are not balanced, the motion of the object will change.

1 Analyse the forces acting on each object below in the 'Situation' column and complete the table.

Situation	Are the forces balanced?	If not balanced, state the direction the unbalanced force acts.	Describe any changes to this object's motion.
(a)	☐ Yes ☐ No	☐ Up ☐ Down ☐ Left ☐ Right	
(b)	☐ Yes ☐ No	☐ Up ☐ Down ☐ Left ☐ Right	
(c)	☐ Yes ☐ No	☐ Up ☐ Down ☐ Left ☐ Right	
(d)	☐ Yes ☐ No	☐ Up ☐ Down ☐ Left ☐ Right	
(e)	☐ Yes ☐ No	☐ Up ☐ Down ☐ Left ☐ Right	
(f)	☐ Yes ☐ No	☐ Up ☐ Down ☐ Left ☐ Right	

Situation	Are the forces balanced?	If not balanced, state the direction the unbalanced force acts.	Describe any changes to this object's motion.
(g)	☐ Yes ☐ No	☐ Up ☐ Down ☐ Left ☐ Right	
(h)	☐ Yes ☐ No	☐ Up ☐ Down ☐ Left ☐ Right	
(i)	☐ Yes ☐ No	☐ Up ☐ Down ☐ Left ☐ Right	
(j)	☐ Yes ☐ No	☐ Up the slide ☐ Down the slide	
(k)	☐ Yes ☐ No	☐ Up ☐ Down ☐ Left ☐ Right	
(l)	☐ Yes ☐ No	☐ Up ☐ Down ☐ Left ☐ Right	

RATE MY UNDERSTANDING
Shade the face that shows your rating   

7.3 Friction around you

Science understanding

FOUNDATION | **STANDARD** | ADVANCED

Have you ever tried to help someone start a car by pushing it? It is very hard to get the car rolling to begin with. It seems to be easier to push once it has started to move. The car tyres have a force working against them. This force works against the forward motion of the car or is said to be resisting the movement. This force is called friction. Friction acts whenever one object moves over another. It always acts in the opposite direction to an object's motion.

Look at the examples of motion below and complete the table. For each example:

1 Identify which two surfaces are in contact.

(2) Indicate if the friction is useful or unwanted and how to lessen the friction if unwanted.

Situation	Surfaces in contact	Friction useful or unwanted?	How to lessen unwanted friction?
filing nails			
riding a skateboard			
going down a waterslide			
pushing a refrigerator			
writing with a pencil			

RATE MY UNDERSTANDING
Shade the face that shows your rating

7.4 Measuring friction

Science understanding

FOUNDATION | **STANDARD** | ADVANCED

Chang and Amalia completed an experiment to see whether adding mass to a tray affected the size of friction when they tried to slide it along a bench top (Figure 7.4.1). The tray and fold-back clip were measured to have a mass of 100 grams. The students pulled the tray and clip carefully and measured the reading on the spring balance while the tray was sliding. Their results for the experiment are shown in Table 7.4.1.

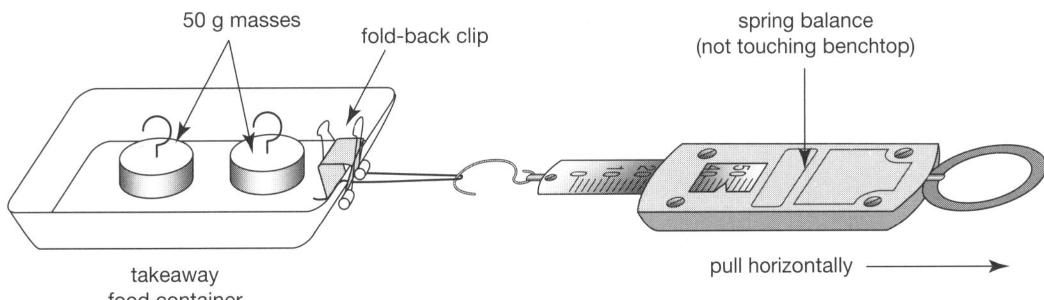

Figure 7.4.1

Table 7.4.1 Results of Chang and Amalia's experiment

Total mass of tray and weights (g)	Sliding friction (N)			Average sliding friction (N)
	Trial 1	Trial 2	Trial 3	
100	0.3	0.5	0.4	
150	0.8	0.8	0.8	
200	1.4	1.5	1.0	
250	1.6	1.9	1.9	
300	2.3	2.2	2.4	
350	2.5	2.5	3.1	
400	3.1	3.4	3.1	
450	3.6	3.7	3.5	

1. Explain why Chang and Amalia measured three readings of sliding friction for each mass tested.

____

____

2. Add the forces measured in Trial 1, Trial 2 and Trial 3 and divide your result by three to calculate the average force measured in each case. Complete the last column of the table. Round each average to one decimal place.

3. Chang and Amalia suspected that they made an error when measuring one result. Identify the likely error by circling this result in the table.

4. Suggest how the students may have recorded an incorrect reading.

5. Plot a graph of results on the axes provided using the average sliding friction.

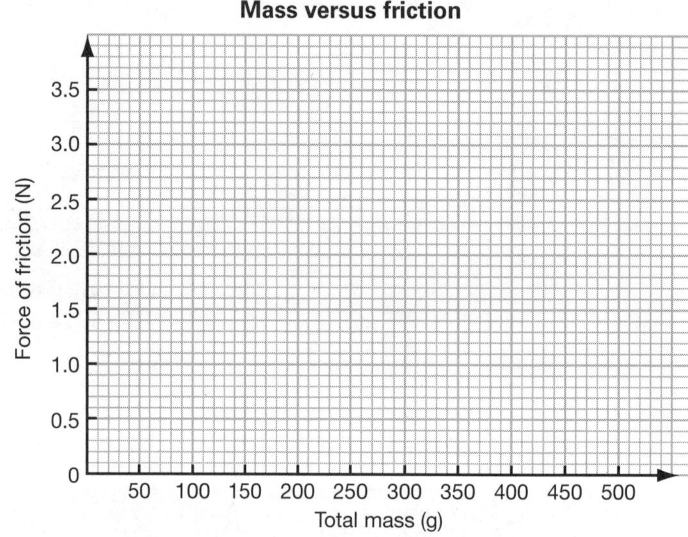

6. **(a)** Insert arrows on the diagram below to show four forces acting on the tray while it is being pulled at a constant speed.

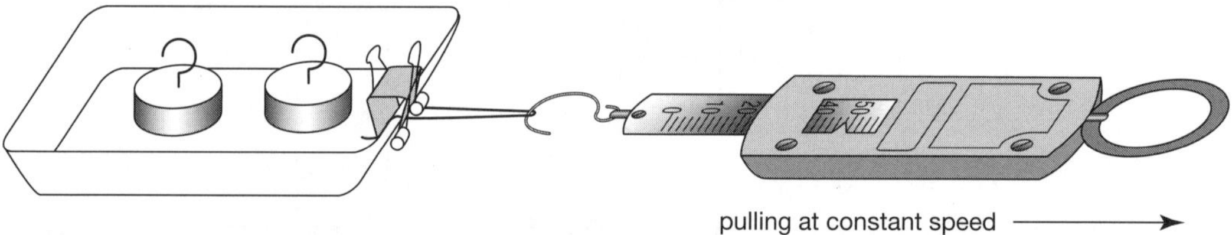

(b) State whether the forces acting on the tray are balanced or unbalanced at this time.

7. What conclusion can you draw from Chang and Amalia's experiment?

8. Describe a real-world example where you have noticed a link between friction and mass.

RATE MY UNDERSTANDING
Shade the face that shows your rating   

Science as a human endeavour

FOUNDATION | STANDARD | **ADVANCED**

Sand playground makes for safer landings

In the often high-tech world of modern children's playgrounds, a new study suggests that an old-fashioned solution may be best.

Canadian researchers have found that children are nearly five times less at risk of breaking arms, legs or parts of their bodies, if their playground has sand on the ground rather than a wood fibre base.

The Toronto team's results are published on the PLoS Medicine website today and may send local councils and planning authorities across Australia scrambling back to their drawing boards.

'We suspect that the fracture rates are lower on sand because of lower surface friction,' write the Toronto-based researchers. 'A lower friction surface allows the hand to slide or sink, limiting the bending moment and preventing a fracture.'

'Updating playground safety standards to reflect this information will reduce the most common and severe injuries seen on modern playgrounds, without limiting children's access to healthy outdoor play.'

Breaking new ground

The researchers describe how they took advantage of a 2003 decision by the Toronto District School Board to resurface the district's school playgrounds. They used the rare opportunity to conduct a randomised controlled trial of the two surfaces.

It found that children playing in those playgrounds with wood fibre surfaces were 4.9 times more likely to suffer fractures than the children playing in sand-covered playgrounds. The rate of fractures reported for the sandy playgrounds was 1.9 for every 100 000 student-months of use.

Australian playground safety expert and associate professor at the University of Technology, Sydney, David Eager says that it is the first such study comparing surfaces in real world conditions and it is very thorough.

'It's an excellent study and long overdue.'

In Australia, a recent study of playground safety found that over a ten-year period there were 106.6 fractures per 100 000 students.

Eager attributes the Australian injury rate in part to the fact that many playgrounds around the country are now surfaced with rubber. His recent research has shown that rather than dispersing the energy generated by a fall, rubber rebounds energy back into the falling body.

'My own opinion is that both sand and wood fibre, or bark, are excellent surfaces for playgrounds and it's the rubber that is the bad surface which causes the rebound and lots more breaks.'

'The forces are significantly greater on the rubber than they are on the bark or the sand.'

Sinking to safe depths

Adequate maintenance of playground surfaces can also play an important role in safety and may have influenced the impressively small injury outcome in the Canadian study.

Any coverings of depth, such as sand or wood fibre, should be maintained at optimum depth to have the desired effect on safety.

While no measurements were made, the researchers point out that the playgrounds were well looked after. They add that not only were the surfaces newly laid, but that the supervisors of the participating playgrounds were aware of the research and worked to maintain the grounds.

Previous research by the NSW Injury Risk Management Research Centre has demonstrated that the safety benefit of such surfaces is significantly diminished if they are not as deep as required. The Centre suggested that it is one area often overlooked during maintenance.

'In Australia, we have lots of old playgrounds and lots of playgrounds that aren't maintained,' says Eager.

'When we do have sand and bark, even if it is put in at the correct depth, at the bottom of slides and climbing poles and other forced movement devices, it all gets pushed away and you're left with something quite hard underneath which is probably where all the injuries are occurring.'

by Annabel McGilvray,
from ABC Science website

1 (a) Researchers from Canada found that children who played in playgrounds with a base of sand were at a much lower risk of breaking a bone in a fall than those who played in playgrounds with a wood fibre base. According to the article, how much greater is the risk of breaking bones on wood fibre than on sand?

____

____

(b) State the reason why the researchers believe this to be the case.

2 If a child falls off playground equipment, explain why their arm or leg is less likely to break if it can slide in the sand.

3 What was the number of fractures over a 10-year period out of a total of 100 000 children in Australia?

4 Safety expert David Eager criticised the use of rubber surfaces in many Australian playgrounds. Outline his reasons for this view.

5 Explain why it is important for playground surfaces to be adequately maintained.

RATE MY UNDERSTANDING
Shade the face that shows your rating

7.6 Measuring pressure

Science understanding

FOUNDATION	STANDARD	ADVANCED

It is hard to walk on soft snow in regular shoes without sinking, but you can walk on soft snow while wearing a pair of wide snow shoes (Figure 7.6.1). This is because your weight force is spread over a wider area. Polar bears have very large feet for this reason. Pressure, P, is the amount of force per unit area. It is measured in a unit called the pascal (Pa) and can be calculated:

regular shoes

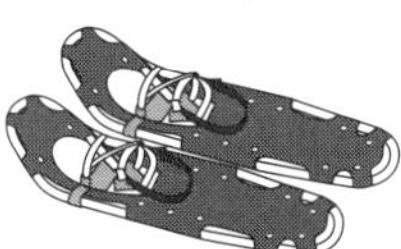

snow shoes

Figure 7.6.1

$$P = \frac{F}{A} \text{ where } F \text{ is force (N) and } A \text{ is area (m}^2\text{)}$$

1 Calculate the pressure you exert on the ground when standing in bare feet.

(a) On the next page trace the outline of your left foot on the grid paper.

(b) Count the number of 1 cm squares contained within the outline of your foot. (Estimate and add part squares to the number of whole squares.)

Area of left foot = ________ cm^2

To convert this value into square metres, divide it by 10 000 (or move the decimal point four places to the left).

Area of left foot = ________ m^2

Assuming your feet are the same size, multiply this result by 2.

Area of both feet = ________ m^2

(c) Calculate the size of your weight force by multiplying your mass (in kilograms) by 10 (approximate gravitational field strength).

My weight force = ________ N

(d) Calculate the pressure you exert on the floor by dividing your weight (in newtons) by the area of both feet (in m^2). (Note: Round your answer to one decimal place.)

$$P = \frac{F}{A} = \frac{\text{weight force (N)}}{\text{area of both feet (m}^2\text{)}} = \_\_\_\_\_\_\_\_ = \_\_\_\_\_\_\_\_ \text{ Pa}$$

2 Predict whether the pressure you exert on the ground would increase or decrease when you are standing:

(a) in a pair of slippers a few sizes larger than your own ________________

(b) on the tip of your toes ________________

(c) in a pair of spiked football boots. ________________

3 Propose why a drawing pin can be pushed easily into a corkboard, but a coin of the same mass and pushed with the same force cannot.

____

4 Analyse why high-heeled shoes, particularly stilettos, can damage a wooden floor.

____

RATE MY UNDERSTANDING
Shade the face that shows your rating

7.7 Buoyancy

Science understanding

FOUNDATION | STANDARD | **ADVANCED**

account (*v*) to make allowance for

cargo (*n*) the goods carried by a ship, plane or train

exert (*v*) to use force or power

inflated (*adj*) full of air

If you try to push an inflated balloon under water, it will keep popping back up. You can feel a force pushing the balloon upwards as you try to force it down. This is because the water exerts an upwards force on the balloon. This force is called buoyancy. The Greek mathematician Archimedes discovered that the size of this force is equal to the weight of water that has been displaced (pushed aside) by the object.

Whether something sinks or floats depends on its density. Density is a measure of the mass that is packed into a volume.

Consider the steel ball shown in Figure 7.7.1. As this ball has quite a small volume, it displaces a small weight of the water. The upwards buoyancy force is less than the gravity pulling the ball down, so it sinks. In contrast, the hull of a ship has a larger volume and so displaces a large weight of water. This produces a large buoyancy force (greater than the weight of the ship), and so the ship floats. If the ship was made from solid steel then its weight force would be greater and it would sink. In other words, if the ship is of lower density (mass per volume) than the density of the water, then it will float. If more dense, then it will sink.

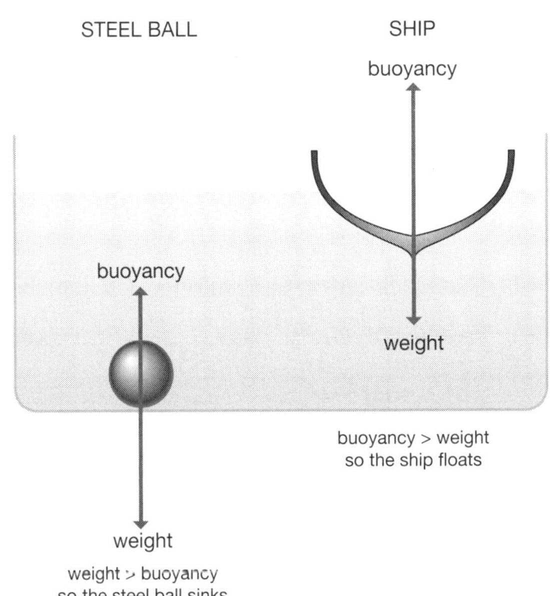

Figure 7.7.1 The relative sizes of the upwards buoyancy force and downwards weight force result in the steel ball sinking and the ship staying afloat.

Sea water has a greater density than fresh water. This means that sea water gives you more buoyancy than the fresh water of a lake or the local pool. The Dead Sea, located on the Israel–Jordan–Palestine border, is one of the saltiest bodies of water found on Earth. As a result, it exerts such a large buoyancy force that it is impossible for a person to sink!

Cargo ships

Ships that carry cargo around the world will travel through waters of varying saltiness. Some ports may be less salty due to the inflow of fresh water from rivers. For this reason, all large ships have a set of markings on their hulls. These markings show the safe level that the ship can be pushed into the water with its cargo to account for buoyancy in different waters. This is shown in Figure 7.7.2.

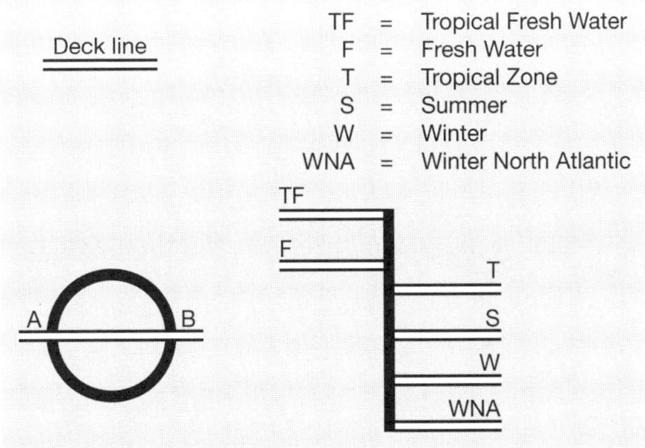

Figure 7.7.2 International Load Lines or Plimsoll Lines for ships

7.7 Buoyancy

(1) Name the force that acts upwards on a rubber duck floating in a bath.

____

(2) Using your knowledge about buoyancy, choose the correct alternative to complete each sentence.

(a) If the buoyant force on an object is greater than its weight force, then the object will float / sink.

(b) If the buoyant force on an object is less than its weight force, then the object will float / sink.

3 State whether:

(a) sea water has a greater or lesser density than fresh water ______________

(b) the buoyancy force of salt water is greater or less than that of fresh water. ______________

(4) Assess whether the following objects would float or sink in water.

(a) a leaf ______________

(b) a marble ______________

(c) a sheet of paper ______________

(d) an apple ______________

(e) a football ______________

(f) a padlock ______________

(g) a cork ______________

(h) a plastic tumbler ______________

(5) A ship can be described as a thin shell with lots of air inside. Explain how this helps the ship to float.

____

____

6 **(a)** Suggest the method of an experiment that could be conducted to compare the buoyancy of objects in water to those in olive oil.

____

____

(b) Name the variables in the experiment.

____

____

RATE MY UNDERSTANDING
Shade the face that shows your rating   

7.8 Static electricity

Science understanding

FOUNDATION | STANDARD | **ADVANCED**

static electricity (*n*) electric charge created when two objects rub together and electrons move from one object to another

The list on the right forms part of the triboelectric series. This lists materials in order of how easily they lose electrons. Materials above cotton tend to lose electrons easily, and may become positively charged. Materials below cotton readily gain electrons and in doing so become negatively charged.

Triboelectric series
- air
- human hands
- asbestos
- glass
- human hair
- nylon
- wool
- fur
- lead
- silk
- aluminium
- paper
- cotton
- steel
- wood
- hard rubber
- nickel, copper
- silver
- gold
- acetate
- polyester
- polystyrene (styrofoam)
- polyethene
- vinyl
- Teflon

1 State whether air is likely to lose or gain negative charges.

____

(2) A plastic acetate sheet is rubbed against your hair and your hair is attracted to it.

(a) State which has become positively charged. ____________

(b) State which has become negatively charged. ____________

(c) Explain why your hair is attracted to the acetate. ____________

(3) When styrofoam is charged with a piece of silk, identify which material loses electrons.

____

(4) On a warm and dry day, the fur of Freddie the mouse is seen to be attracted to Josef's vinyl shoes. Explain what is happening in this situation.

____

____

(5) Use these diagrams to state the number of positive and negative charges in each atom, and classify each as having a positive charge, negative charge or no overall charge.

(a)

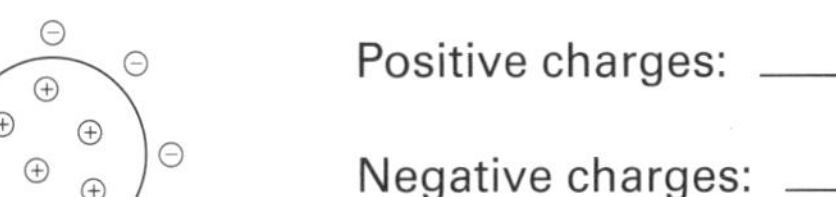

Positive charges: ____________

Negative charges: ____________

Overall charge: ____________

(b)

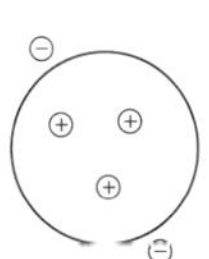

Positive charges: ____________

Negative charges: ____________

Overall charge: ____________

HINT
Calculate the overall charge by adding together the positive charge and negative charge.

RATE MY UNDERSTANDING
Shade the face that shows your rating

7.9 Understanding levers

Science understanding

FOUNDATION | **STANDARD** | ADVANCED

Levers are simple machines. A lever pivots about a point called the fulcrum. An effort force is applied to a lever so that it can work against a load.

There are three classes of lever. First-class and second-class levers are force multipliers and increase the force you can apply. Third-class levers actually have a force disadvantage, meaning that you need to supply a greater effort force for the job than would be needed without the lever. However, these levers provide a distance or speed advantage.

The positions of the effort force (E), fulcrum (F) and load force (L) are summarised below.

1 Study each item shown below in images (a) to (b). Label where you think the effort force (E), fulcrum (F) and load force (L) are positioned on each diagram and then classify each as a first-class, second-class or third-class lever. Labeled diagrams of first-, second- and third-class levers are provided in the hints box.

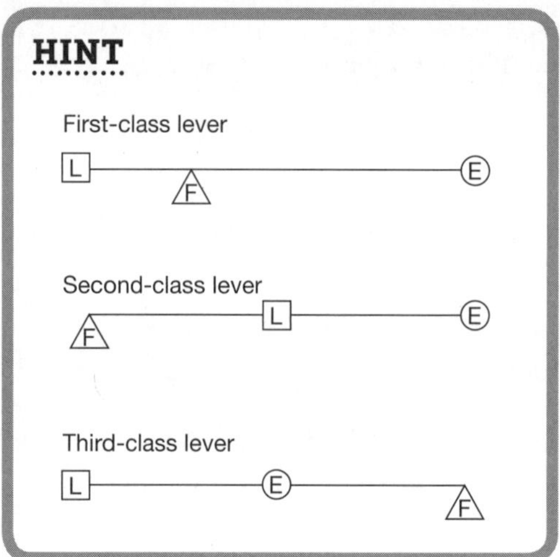

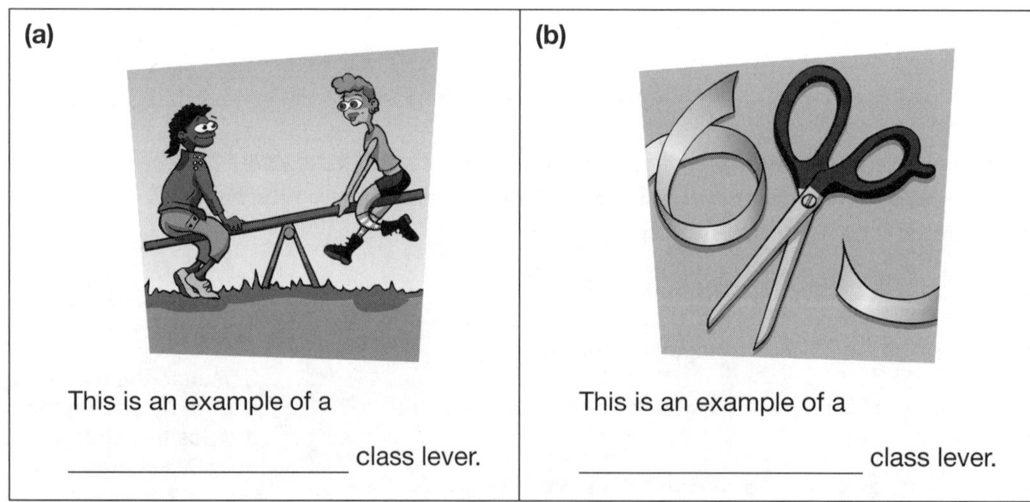

(c)

This is an example of a

__________________ class lever.

(d)

This is an example of a

__________________ class lever.

(e)

This is an example of a

__________________ class lever.

(f)

This is an example of a

__________________ class lever.

(g)

This is an example of a

__________________ class lever.

(h)

This is an example of a

__________________ class lever.

2 List which levers are force multipliers.

____

____

3 List which levers are distance or speed multipliers.

____

____

RATE MY UNDERSTANDING
Shade the face that shows your rating 

7.10 Classifying machines

Science understanding

FOUNDATION | **STANDARD** | ADVANCED

1 Identify one or more simple machines in each situation below. Select from: lever, gears, inclined plane, pulley, wedge, screw. Record these under each diagram.

2 Classify which devices increase the force applied and which increase speed. Circle devices that multiply force in blue. Circle those that multiply speed in red.

(a)

____

(b)

____

(c)

____

(d)

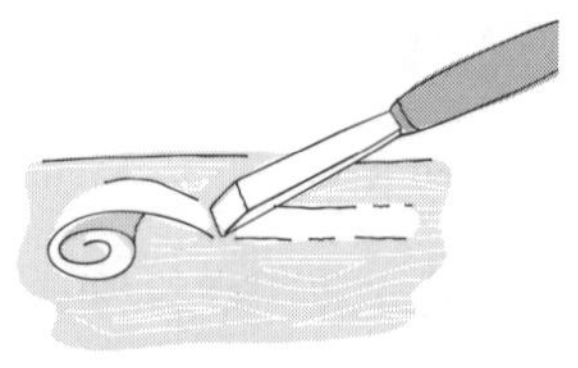

____

(e)

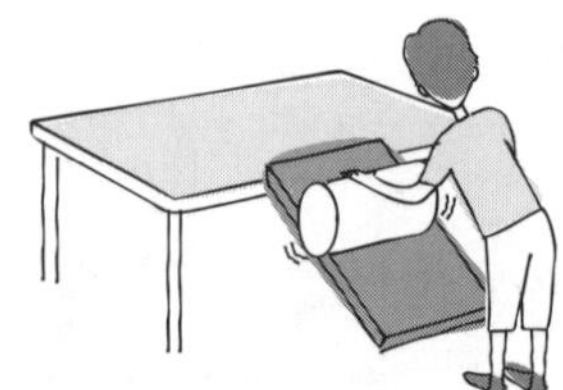

____

(f)

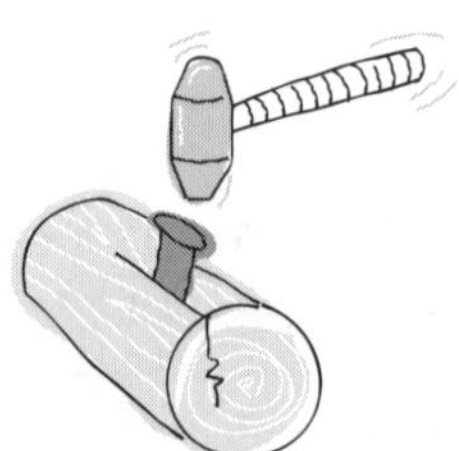

____

(g)

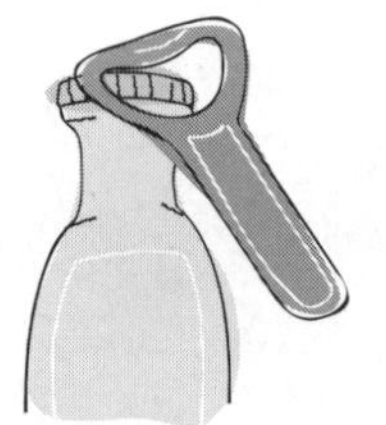

____

(h)

____

(i)

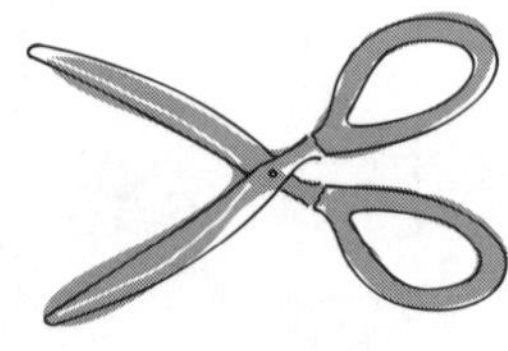

____

(j)

____

(k)

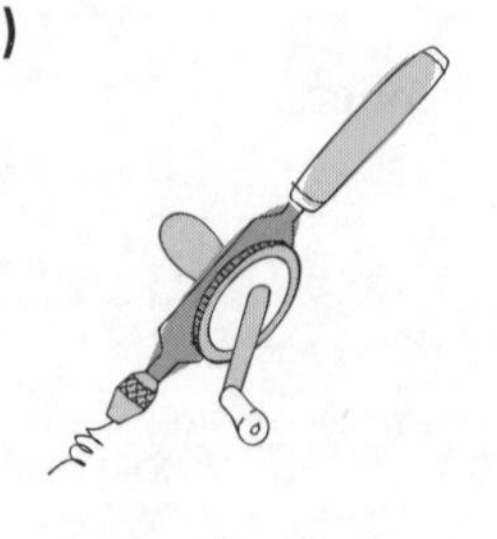

____

(l)

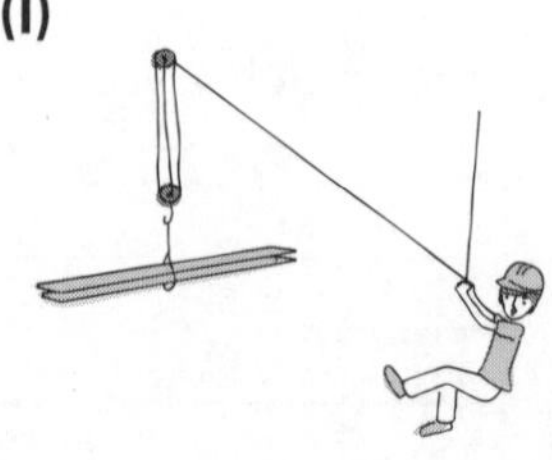

____

RATE MY UNDERSTANDING
Shade the face that shows your rating  

7.11 Literacy review

Science understanding

FOUNDATION	STANDARD	ADVANCED

Recall your knowledge of forces by correctly inserting words from the list below to complete the statements. Use each word once.

attract	direction	effort	field	friction
fulcrum	load	mass	newton	pull
repel	size	speed	surface	weight

1 A force can be a push, a ______________________ or a twist.

2 A force is measured using a spring balance by using a unit called the ______________________ .

3 When pushing a refrigerator across the floor, you need to overcome the force of ______________________ . This force exists whenever one ______________________ slides over another.

4 Your ______________________ is a downwards force due to gravity.

5 ______________________ is the amount of matter in a substance.

6 Like magnetic poles will ______________________ while unlike magnetic poles will ______________________ .

7 The region around an object in which a non-contact force can be experienced is called a ______________________ .

8 A simple machine can increase the ________________ of a force, the ________________ of a force or the ________________ something moves.

9 The force that you apply to a machine is called the ______________________ .

10 The force that a machine must overcome is called the ______________________ .

11 A lever pivots about a point called the ______________________ .

RATE MY UNDERSTANDING
Shade the face that shows your rating

7.12 Thinking about my learning

Complete this review thoughtfully and carefully, and write your review in complete sentences and paragraphs.

1 What have you been learning about in the topic on forces?

____

____

____

____

____

____

____

____

2 Four important things I learnt in the unit on forces are:

(i) ____

(ii) ____

(iii) ____

(iv) ____

3 Two things I am very proud of achieving in this unit on forces are:

(i) ____

(ii) ____

4 Two things that I need to improve are:

(i) ____

(ii) ____

5 On a scale of 1 to 10, where 1 is unsatisfactory and 10 is excellent; rate the following by placing a red dot on each scale.

(i) My understanding of the content. 0 – 10

(ii) My skills in doing practical investigations and writing reports. 0 – 10

(iii) The effort I have put into this topic. 0 – 10

CHAPTER 8

Earth in space

8.1 Knowledge preview

Science understanding

FOUNDATION | STANDARD | ADVANCED

1 Test what you know about space and astronomy yourself by completing the chart below.

Key terms that you know:

Fact 1

Earth in space

Astronomy

Fact 2

Fact 3

Fact 4

Fact 5

Fact 6

Fact 7

Fact 8

8.2 Using sky maps

Science understanding

FOUNDATION | **STANDARD** | ADVANCED

constellation (*n*) a formation of stars that might look like an object, a person or an animal

Milky Way (*n*) the galaxy that Earth and our solar system are part of. It is called 'milky' because the light from the millions of stars looks like a white band across the sky.

A sky map (or planisphere) is a map of the night sky. The circumference or outer edge of the map represents the horizon. The Milky Way is represented by the shaded band running across the sky map. Stars are represented by dots: the brighter the star, the bigger the dot. Many stars are connected by lines to indicate common constellations.

The apparent position of stars in the sky changes over the year, so each sky map needs to be drawn to suit the date, time and location on the Earth from where you will be viewing the sky. You can go to http://www.skymaps.com to print a sky map for the month. (Make sure you download the southern hemisphere edition.)

HINT

When we look at a road map, we use the direction indicators north, south, east and west. Because a sky map is 'upside down', east and west are reversed.

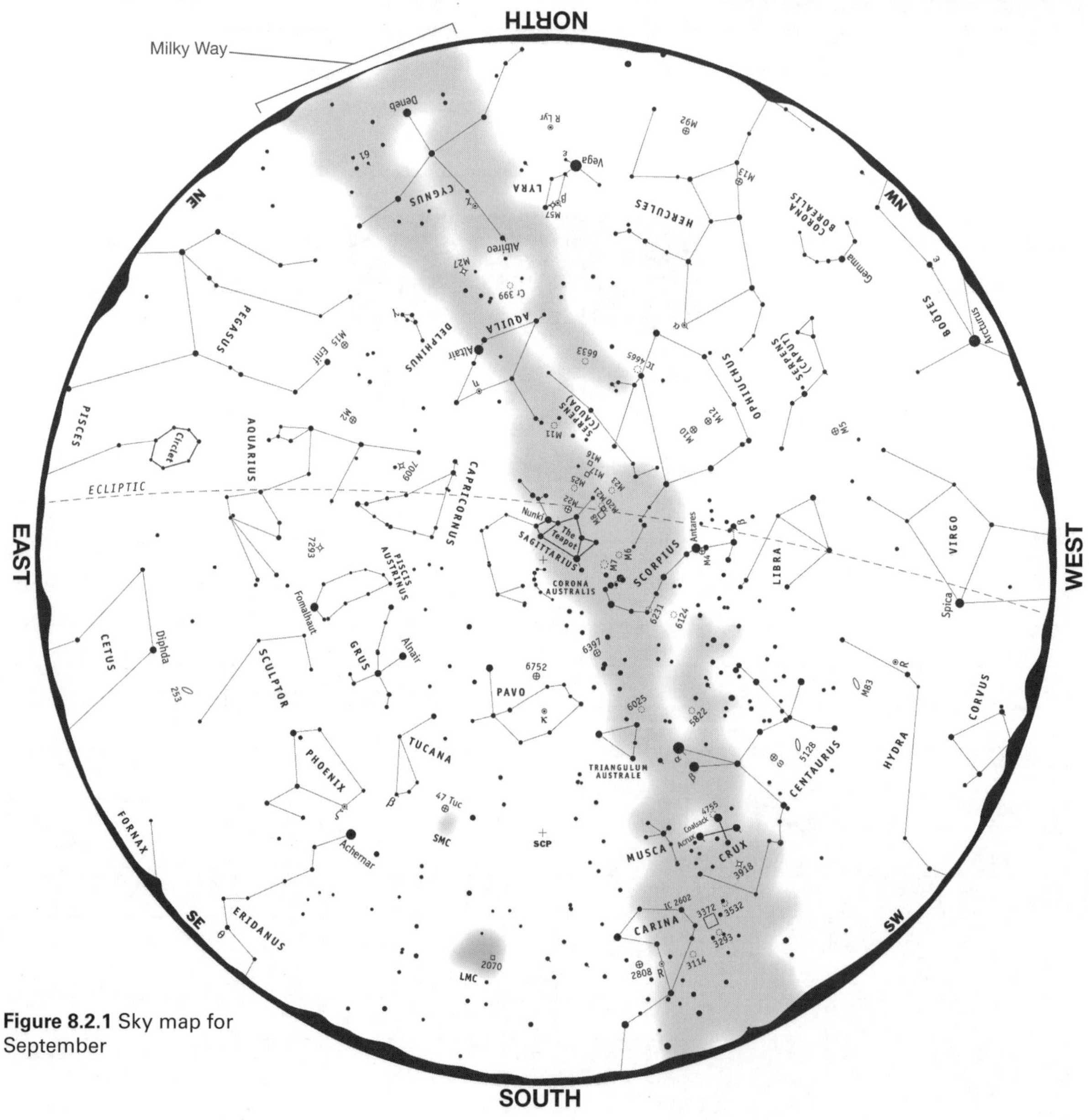

Figure 8.2.1 Sky map for September

8.2 Using sky maps

Refer to the sky map in Figure 8.2.1 to answer the following questions about astronomical observations made on a night in September.

> **HINT**
> The Southern Cross (Crux) is a small constellation visible in the southern hemisphere. It contains five bright stars. The four brightest stars make up a cross that points towards the South Celestial Pole (which is marked SCP on the sky map). The Southern Cross appears on the Australian flag.

1 State in which direction an observer should face to observe the Southern Cross (Crux).

____

2 Name the constellation that can be seen on the western horizon.

____

3 Highlight which of the following best describes the direction that the Milky Way runs.

A north–south

B east–west

C south–east–north–west

4 Name the brightest star in the constellation Piscis Austrinus. (Look in the eastern section of the sky map.)

____

(5) The South Celestial Pole is indicated by its initials (SCP). Find this on the sky map.

An old astronomical technique for finding the South Celestial Pole in the sky is explained below.

- Draw a line through the long arm of the Southern Cross. Extend to the left.
- Draw a perpendicular bisector between the Pointers. To do this, follow the steps below:
  - Find the two Pointers (i.e. α-Centauri and β-Centauri. These are the two brightest stars in the constellation Centaurus).
  - Draw a connecting line between these two stars.
  - Find the mid-point of the connecting line.
  - Starting at the mid-point of the connecting line, draw a line at right angles to the connecting line towards the centre of the sky map. This is called a perpendicular bisector.
- Find where the line through the Southern Cross and the perpendicular bisector cross. This should be the South Celestial Pole.

(a) Apply this technique on the sky map.

(b) Describe whether it worked.

____

(c) Propose how the task is different when carried out on a flat sky map compared to following these instructions when looking at the sky itself.

____

____

RATE MY UNDERSTANDING Shade the face that shows your rating			☹

8.3 Astrology investigated

Science as a human endeavour

FOUNDATION	STANDARD	ADVANCED

astrology (*n*) the belief that the position of stars and planets has an influence on human affairs

The table below displays the names of star signs according to a person's birth date.

Sign	Represents	Born
Aries	Ram	21 March–19 April
Taurus	Bull	20 April–20 May
Gemini	Twins	21 May–20 June
Cancer	Crab	21 June–22 July
Leo	Lion	23 July–22 August
Virgo	Virgin	23 August–22 September
Libra	Scales	23 September–22 October
Scorpio	Scorpion	23 October–21 November
Sagittarius	Archer	22 November–21 December
Capricorn	Goat	22 December–19 January
Aquarius	Water bearer	20 January–18 February
Pisces	Fish	19 February–20 March

1 Find your star sign according to the table. If possible, find at least one other person with the same star sign and complete the rest of this task together.

2 Read through the personality traits listed below. Identify the two words that best describe you.

Star sign personality traits

ambitious	careful	caring	challenging
communicative	conservative	curious	decisive
dominant	fair	imaginative	impatient
independent	negotiating	orderly	original
passionate	playful	proud	reliable
searching	secretive	sensitive	thoughtful

I am ________________________ and ________________________.

8.3 Astrology investigated

Table 8.3.1 shows the links between these personality traits and the various star signs according to one astrology website.

Table 8.3.1 Astrology and related personality types

Sign	Personality traits
Aries	challenging, curious
Taurus	conservative, reliable
Gemini	communicative, playful
Cancer	caring, sensitive
Leo	dominant, proud
Virgo	careful, orderly
Libra	negotiating, fair
Scorpio	secretive, passionate
Sagittarius	impatient, independent
Capricorn	ambitious, decisive
Aquarius	thoughtful, original
Pisces	searching, imaginative

3 Do the words you chose match the description given by astrologers?

____

____

4 Identify which star sign's description best fits your personality.

____

____

____

RATE MY UNDERSTANDING
Shade the face that shows your rating

8.4 Tides

Science inquiry skills

FOUNDATION	STANDARD	ADVANCED

The times for high and low tides for locations around Port Phillip Bay in Victoria are shown in Table 8.4.1. The alignment of the moon and the sun with the Earth is the main cause of high and low tides. The gravitational attraction of the moon pulls oceans out towards the moon causing a low tide.

Table 8.4.1 Port Phillip Bay high and low tide times

Date	Location	High tide	Low tide
16 July	Williamstown	8:36 a.m. 8:23 p.m.	2:36 a.m. 2:10 p.m.
	Port Phillip Heads	5:29 a.m. 5:30 p.m.	10:57 a.m. 11:29 p.m.
	Stony Point	6:28 a.m. 6:18 p.m.	11:57 a.m.
17 July	Williamstown	9:28 a.m. 8:55 p.m.	3:16 a.m. 2:50 p.m.
	Port Phillip Heads	6:20 a.m. 6:05 p.m.	11:40 a.m.
	Stony Point	7:15 a.m. 6:53 p.m.	12:24 a.m. 12:40 p.m.
18 July	Williamstown	10:28 a.m. 9:31 p.m.	4:01 a.m. 3:37 p.m.
	Port Phillip Heads	7:17 a.m. 6:45 p.m.	12:08 a.m. 12:25 p.m.
	Stony Point	8:08 a.m. 7:32 p.m.	1:01 a.m. 1:26 p.m.

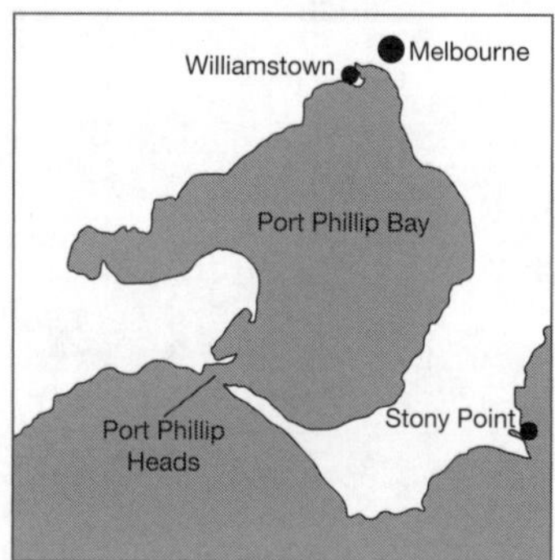

Use Table 8.4.1 to answer the following questions.

1 State how many high tides and how many low tides occur each day at most places around the bay.

____

(2) Propose a reason why low-tide data for Port Phillip Heads and Stony Point seems to be missing.

____

____

____

3 State whether all places on Port Phillip Bay experience tides at the same time.

____

4 Identify whether the tides were getting later or earlier from the beginning to end of July.

____

(5) Calculate how much time passes between:

(a) the first low tide and first high tide at Williamstown on 16 July

____

(b) the first high tide and the next high tide at Port Phillip Heads on 17 July

____

(c) the morning high tide at Williamstown on 17 July and the same tide on 18 July

____

(d) the afternoon low tide at Stony Point on 17 July and the same tide on 18 July.

____

(6) Calculate approximately how long it is between:

(a) high tides on the same day

(b) low tides on the same day.

Use the patterns in the tides in Table 8.4.1 to predict the tide times for 19 July and complete the following table.

Date	Location	High tide	Low tide
19 July	Williamstown		
	Port Phillip Heads		
	Stony Point		

8.5 Moon map

Science inquiry skills

FOUNDATION	STANDARD	ADVANCED

Processing & Analysing

1 On a clear night, go outside and locate the Moon in the sky. If it's not there, then try again later or try to find it the next night. Once you find it, draw a sketch of it and its shape.

2 Look at the following images of the Moon. Circle the one that most closely matches the Moon as it appears tonight. Identify the phase of the Moon (new moon, waxing or waning crescent moon, quarter moon, waxing or waning gibbous moon or full moon).

_______ _______ _______ _______ _______ _______ _______ _______ _______

3 Observe the surface of the Moon with no assistance from binoculars or a telescope. Tick the features that you can identify on Figure 8.5.1.

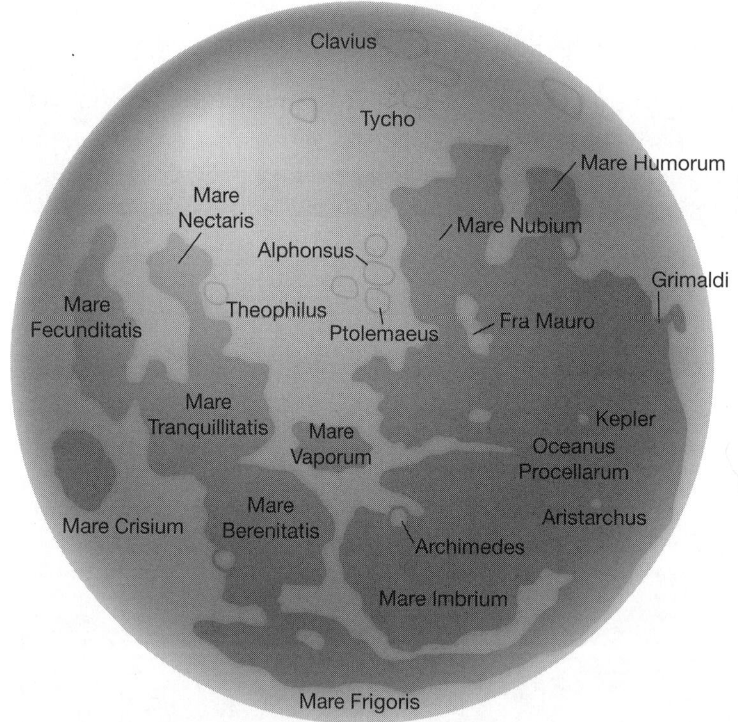

Figure 8.5.1 Map of features of the Moon

4 If you have access to a telescope or binoculars, then use them to study the Moon again. Continue to tick the features that you can identify.

RATE MY UNDERSTANDING
Shade the face that shows your rating   

8.6 Time zones

Science inquiry skills

FOUNDATION | **STANDARD** | ADVANCED

Processing & Analysing

Figure 8.6.1 shows a map of the world with each time zone separated by black lines.

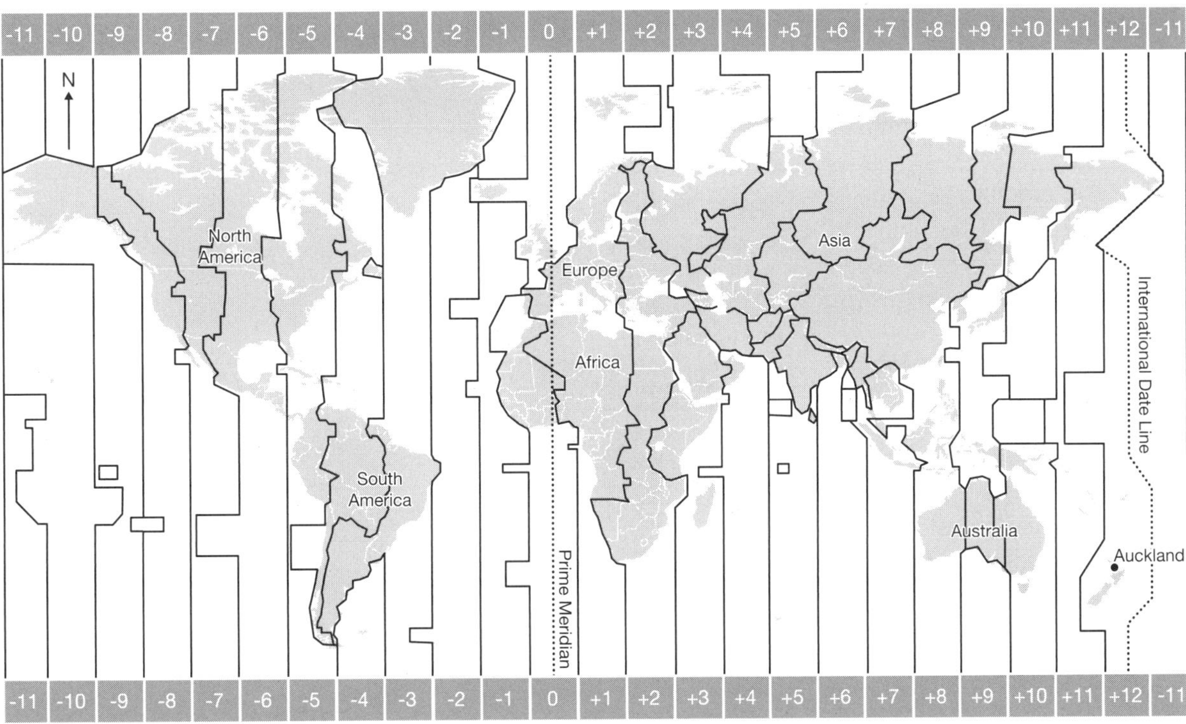

Figure 8.6.1 Map of world time zones

1 State how many times zones are in:

(a) Australia ________

(b) Africa ________

(c) North America (USA and Canada) ________

2 Identify whether the time zones run from north to south or east to west.

____

3 Use the rotation of the Earth to explain your answer to question 2.

____

____

____

8.6 Time zones

Another way of showing time zones is to state what the time is when it is midday in Great Britain (Greenwich Mean Time or UTC), as shown in Table 8.6.1.

Table 8.6.1 Times around the world

Location	Time
Greenwich Mean Time (UTC)	12.00 (noon or midday)
Adelaide, Australia	21.30 (9:30 p.m.)
Auckland, New Zealand	24.00 (midnight)
Brisbane, Australia	22.00 (10:00 p.m.)
Darwin, Australia	21.30 (9:30 p.m.)
Dubai, United Arab Emirates	16.00 (4:00 p.m.)
Hobart, Australia	22.00 (10:00 p.m.)
Honolulu, United States	02.00 (2:00 a.m.)
London, Great Britain	12.00 (midday)
Melbourne, Australia	22.00 (10:00 p.m.)
New York, United States	07.00 (7:00 a.m.)
Perth, Australia	20.00 (8:00 p.m.)
Rome, Italy	13.00 (1:00 p.m.)
Sydney, Australia	22.00 (10:00 p.m.)

4 Refer to Figure 8.6.1 and Table 8.6.1 to identify:

(a) which Australian capital cities are on the same time zone

____

(b) which major city is on the same time zone as UTC.

____

5 It's 10 p.m. (22.00) in Sydney. Calculate the time it is in:

(a) Adelaide ________ **(b)** Rome. ________

6 It's midnight (0.00 or 24.00) in Adelaide. Calculate the time it is in:

(a) Auckland ________ **(b)** Perth. ________

7 It's 8 a.m. (08.00) in Perth. Calculate the time it is in:

(a) Melbourne ________ **(b)** London ________ **(c)** Adelaide. ________

8 It's 2 a.m. (02.00) in London. Calculate the time it is in:

(a) New York ________ **(b)** Dubai ________ **(c)** Perth. ________

RATE MY UNDERSTANDING
Shade the face that shows your rating  

8.7 Your age on other planets

Science inquiry skills

FOUNDATION	STANDARD	ADVANCED

Processing & Analysing

Different planets have years of different length. This means that your age would be different on each planet.

1 State your age in Earth years here. ______

2 Calculate your age as shown in Table 8.7.1 and complete the table. (Give your answer correct to one decimal place.)

Table 8.7.1 Age conversions for different planets

Planet	Calculation required	Your age on the planet
Mercury	multiply your age by 365 then divide by 88	
Venus	multiply your age by 365 then divide by 225	
Earth	do nothing	
Mars	multiply your age by 365 then divide by 687	
Jupiter	divide your age by 12	
Saturn	divide your age by 29.5	
Uranus	divide your age by 84	
Neptune	divide your age by 164.8	

3 State how many birthdays you would have celebrated if you lived on:

(a) Earth ______

(b) Mercury ______

(c) Neptune. ______

RATE MY UNDERSTANDING
Shade the face that shows your rating  

8.8 Literacy review

Science understanding

FOUNDATION | STANDARD | ADVANCED

1 Use the clues below to identify the missing words. The first letter of each word is provided.

Clue	Word
space-based	c
an oval shape	e
force of attraction between masses	g
path around a star, planet or moon	o
rocky body found between Mars and Jupiter	a
different shapes of the Moon as seen from Earth	p
mass in orbit around another mass	s
the head of a comet	c
the line connecting the north pole with the south. Planets rotate around it.	a
the time a planet takes to rotate completely once on its own axis	d
caused by the tilt of a planet's axis	s
the time a planet takes to revolve once around the Sun	y
Pluto is one of these	d
shooting star	m
a pattern of stars	c
Earth-like, rocky	t
the gas on a star that is burning	h

2 Once you have found the missing words in question 1, find them and highlight them in the word search puzzle below.

C	X	E	Q	G	S	P	N	L	C	D	T	D	S	C
B	O	Z	T	E	R	E	D	E	O	W	E	I	E	G
F	L	N	D	I	G	A	L	Y	A	D	R	O	A	T
C	R	I	S	O	L	E	V	G	R	K	R	R	S	Z
H	T	A	R	T	S	L	V	I	I	J	E	E	O	T
P	X	D	W	T	E	N	E	D	T	G	S	T	N	T
T	Y	K	I	D	S	L	T	T	A	Y	T	S	S	U
H	P	A	T	F	Z	O	L	M	A	X	R	A	I	F
D	L	O	R	B	I	T	W	A	O	S	I	A	Q	D
D	I	O	R	O	E	T	E	M	T	F	A	S	F	Z
E	L	L	I	P	S	E	Y	V	U	I	L	J	O	I
U	N	N	N	G	C	E	D	Z	L	L	O	D	C	D
I	R	R	V	I	A	R	B	X	P	T	I	N	O	O
D	R	O	C	R	X	I	X	G	V	A	O	P	M	J
S	E	S	A	H	P	F	J	K	C	O	A	D	A	Z

3 Select any eight words from the word search in question 2. Write a short paragraph about space using your eight words. Make sure the paragraph includes a topic sentence and that the sentences that follow all include information that explains the topic sentence.

RATE MY UNDERSTANDING
Shade the face that shows your rating

8.9 Thinking about my learning

Complete the following flow chart to reflect on what you have learnt in this unit of work on Earth in space.

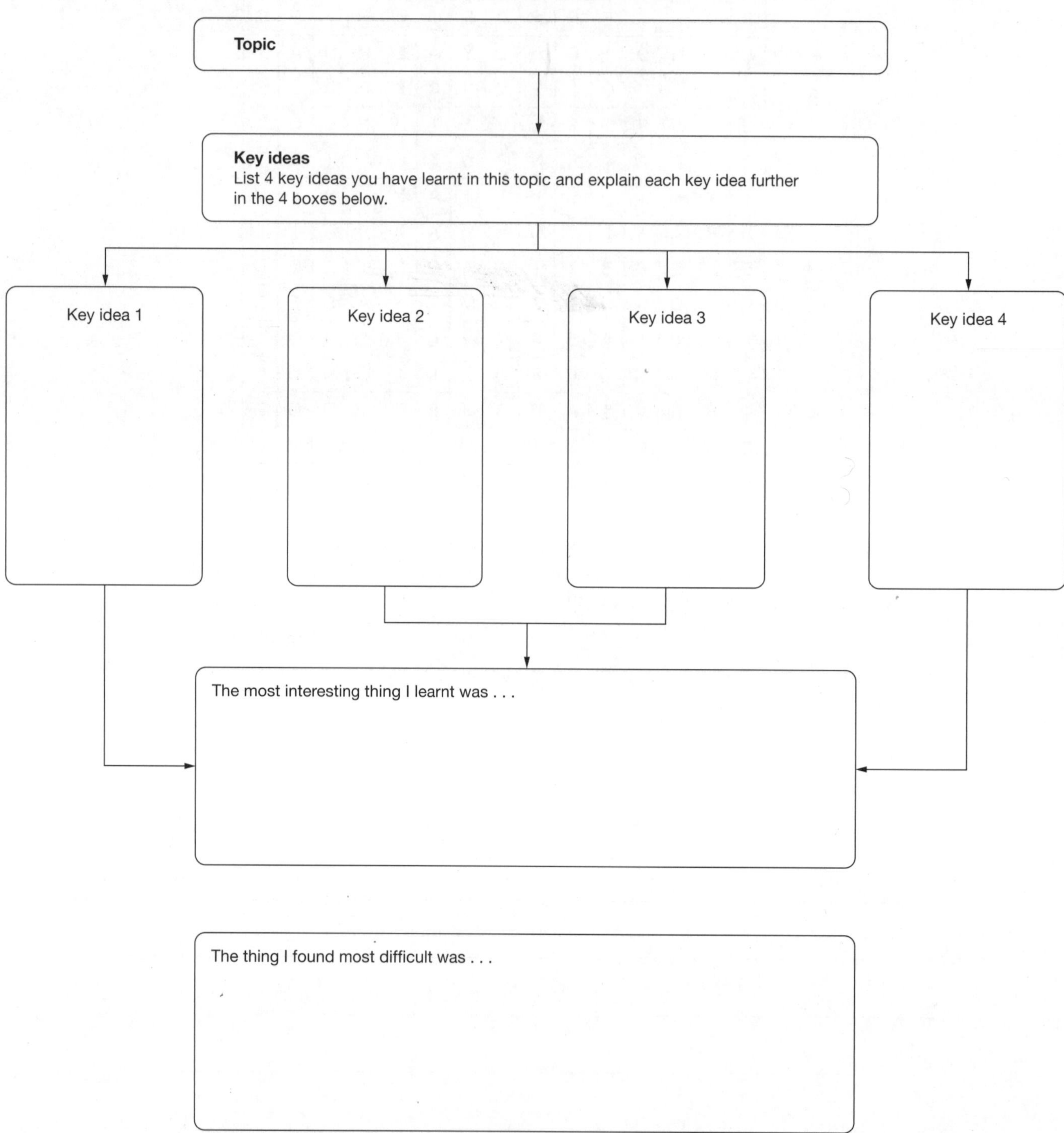

The thing I found most difficult was . . .